HEAT EXCHANGER ENGINEERING TECHNIQUES

PROCESS, AIR CONDITIONING, AND ELECTRONIC SYSTEMS

A Treatise on Heat Exchanger Installations That Did Not Meet Performance

Michael J. Nee

ASME Press
The American Society of Mechanical Engineers
New York, New York

Three Park Ave., New York, NY 10016

Library of Congress Cataloging-in-Publication Data

Nee, Michael J., 1924-
Heat exchanger engineering techniques: process, air conditioning, and electronic systems/Michael J. Nee
p. cm.
Includes bibliographical references and index.
ISBN 0-7918-0167-5
1. Heat exchangers. 2. Cooling. 3. Industrial buildings—Air conditioning. I. Title.

TH7684.F2N44 2003
621.402'5—dc21 2003040423

To My Wife, Virginia, and to Our Children,
Michael, Mary, Rita, and Tim

Engineering

It is a great profession. There is the fascination of watching a figment of the imagination emerge through the aid of science to a plan on paper. Then it moves to realization in stone or metal or energy. Then it brings jobs and homes to men. Then it elevates the standards of living and adds to the comforts of life. That is the engineer's high privilege.

The great liability of the engineer compared to men of other professions is that his works are out in the open where all can see them. His acts, step by step, are in hard substance. He cannot bury his mistakes in the grave like the doctors. He cannot argue them into thin air or blame the judge like the lawyers. He cannot, like the architects, cover his failures with trees and vines. He cannot, like the politicians, screen his shortcomings by blaming his opponents and hope the people will forget. The engineer simply cannot deny he did it. If his works do not work, he is damned. . . .

On the other hand, unlike the doctor his is not a life among the weak. Unlike the soldier, destruction is not his purpose. Unlike the lawyer, quarrels are not his daily bread. To the engineer falls the job of clothing the bare bones of science with life, comfort, and hope. No doubt as years go by the people forget which engineer did it, if they ever knew. Or some politician puts his name on it. Or they credit it to some promoter who used other people's money. . . . But the engineer himself looks back at the unending stream of goodness which flows from his successes with satisfactions that few professions may know. And the verdict of his fellow professionals is all the accolade he wants.

Herbert Hoover

Contents

Preface

The techniques of selecting heat transfer equipment for process, power, industry, air conditioning, and cooling of electronics are mature ones. Plenty of data about these are available in the literature. Professional organizations, industrial companies, and equipment manufacturers have tested many kinds of equipment including upgraded designs to improve product performance. Engineering societies have developed standards, data sheets, and construction details for various designs. To help in making a selection, the thermal properties of most fluids are known or are available. Computer programs have been developed to improve the quality and speed of the selection process. Industry has, in general, reached the point where it is rare that an inadequate heat exchanger is selected. It is not the intent of this writing to challenge the existing data and methodology in any way.

Once a unit has been selected it must be built, assembled, shipped, and installed. At the same time it must meet other criteria such as energy usage, noise, ease of cleaning, weight, flow distribution, and others. It is these criteria that this book addresses. Many examples given, particularly those of requirements omitted from specifications, are well known to some construction companies, as they have faced these kinds of problems previously. These problems are included in this work because they are not common knowledge in all companies.

One purpose of this book is to inform users of conditions that have caused failures on other projects and to give guidance on how they can be avoided on future projects—thus, to assist users in making better selections and to inform them of the limits of other designs. The contents are a cross section of problems that have occurred during the author's 46-year career.

Some points require clarification. The author's experience is that roughly 1 selection in 25 is a poor choice in terms of construction. Further, the correction will usually not be the best solution but only the expedient one.

Few engineers possess the skills to rate all types of heat exchangers. The best design and construction is apt to be overlooked. Also, locating the source of a problem, once it occurs, is not easy and is far more time-consuming, on average, than sizing a unit. Locating the source of a problem usually cannot be done sitting behind a desk.

Even though most exchangers selected are thermally correct, many have not been the best choice for an application. This has resulted in field problems of one kind or another. The intent is to identify problems that occur and suggest ways of preventing their happening again. Hence, this effort addresses such factors as shipping, handling, installation, and servicing that can be simplified by thinking through the various aspects of a problem. To do this, information has been gathered in this book to assist personnel when facing problems similar to these:

1. Nonperformance is often caused by the omission of a requirement from the specifications. Examples are given.
2. Ways of altering an exchanger's geometry to make better use of space are given.
3. Ways to improve handling, maintenance, and cost are identified.
4. Conditions that reduce an exchanger's thermal capacity are described. Hints are provided on how to locate the source of the problem.
5. Reasons are given why some constructions should be avoided on certain applications.
6. A primary goal is to aid personnel (rater, checker, inspector, salesperson) in identifying requirements. Several ways of doing this are included.

The effort covers shell-and-tube units, air coolers, double-pipe units, and plate exchangers. Few manufacturers offer all these products. A good recommendation takes into account these factors:

- Is it a new application?
- Is this a plant expansion where your company was the original supplier?
- Is it an expansion where your company was not the original supplier?
- Is this a new system designed to perform with an existing one?
- Is this a failure situation requiring correction?

A different selection or construction may be recommended in each case. The original specifications are not always available when modifications must be made. Time is a factor that affects a selection. In general, suppliers respond to requests for proposal within three weeks; in failure situations, answers are needed at once. The recommendation varies with the application. To systematize this array of conditions, subjects that in any way apply to process exchangers are addressed in Part 1 of this book, "Process Exchangers." Should a subject apply to air-conditioning (A/C) systems or electronics cooling, the applicable paragraphs are referenced in Part 2 or Part 3, respectively. An effort has been made to cross-reference subjects. In troubleshooting, the assumption will be that exchangers are thermally correct. The exception is undersized A/C evaporators, because these have a strong effect on the operation of the system, prompting personnel to begin complaining about high temperatures almost immediately.

Considerations other than balancing equations may be necessary to determine the best exchanger for an application. These include weight, dimensions, noise level, duct size, relative humidity, filters, and materials, to name a few. This effort is intended to address conditions that influence the design. It alerts readers to designs that have been successful in one industry and that might be useful in another. Factors such as access to the site, lift capacity, space limits, availability of materials, cost, and others ideally should be known before the selection process begins. The end result should be to reduce cost and improve performance. The more completely the requirements are specified, the easier it is to select equipment to meet needs. Extreme or unusual conditions, particularly start-up and shutdown procedures, should be in the specifications. For this purpose information has been assembled in this book to aid in the selection process. Ideally it will help avoid failures of the types that have occurred on other projects. These are identified and should be of use to the specification writer in identifying requirements.

Workable exchangers are the goal. Before a recommendation is finalized, the way the exchanger is to be handled and shipped should be determined. This is true of large items and of the 1 in 10 applications that are modifications to existing facilities and which generally require a minimal turnaround time. Exchangers can be modified to fit the available space more readily than other mechanical equipment. It is difficult, costly, and often impractical to modify exchangers once they are built.

This book includes techniques in exchanger selection, in choosing alternates, and in construction and installation matters. The examples are of failures, errors, or troubles caused mainly by incomplete procedures and/or an omission of requirements which has resulted in field problems. Rarely is a poor thermal selection the cause of failure, although we might often see that another choice would have been better. The information omitted from the specifications would nearly always have influenced the selection, construction, and recommendation and avoided the failure(s) that followed. It should be understood that a given manufacturer may not produce the best surface or construction for an application. The user or the user's agent is in the best position to evaluate the designs and surfaces offered.

Cooling electronics with air is the exception. Heat exchangers are often not needed other than those that are part of an air-conditioning system. Seldom do specifications exist for complete systems; more often they exist for its components. Frequently occurring problems are hot spots, equipment overheating, air recirculation, air distribution, loss of power (circuit breaker trips), condensation, start-up and shutdown concerns, and high relative humidities. Other equipment designed into these systems includes screen rooms, computer room flooring, racks, ducting, EMI/RFI filters, and products associated with noise reduction, humidification, and dehumidification. All affect the A/C system.

This book assumes the reader has some knowledge of how to size heat transfer equipment or can obtain the information from reputable manufacturers or sources. The purpose is to build on this knowledge and identify for user, manufacturer, and student other kinds of information needed to make a good selection. The book also includes conditions to avoid in order to have smoother operating, more economical, and near trouble-free installations.

Michael J. Nee
January 2003

Acknowledgment

Many people provided assistance in the writing of this book. I would like to thank Christine Poulos for providing the computer drawings in a timely fashion usually on short notice. My editor, Mary Grace Stefanchik, provided wise and stimulating encouragement. One could not ask for a more thorough production coordinator than Ray Ramonas who provided expert guidance and sustained accuracy with a lively spirit easing the effort and keeping the book moving. A special honor is due Richard K. Mickey for his excellent suggestions in the editing of this book. I would like to extend a warm thanks to each of my children, Michael, Mary, Rita, and Tim and particularly my wife, Virginia, for their generous encouragement during this long project.

List of Figures

List of Tables

PART 1
PROCESS EXCHANGERS

Chapter 1

Personnel and Requirements

1.1 THE SELECTION OF EXCHANGERS

The heat exchanger engineer's work includes selecting equipment for numerous kinds of applications or to accommodate various conditions:

1. For *new installations,* the task is to select some basic equipment to provide heating or cooling. In this instance, the engineer chooses among shell-and-tube units, air coolers, double-pipe units, plate exchangers, and other designs.

2. The exchanger's size may have to be increased to accommodate a *plant expansion*. New units in series or parallel may be needed to meet the new conditions. Other options are:

- **a.** Replace a shell-and-tube bundle with another having a different size tube, bare or finned, on a new spacing. This usually requires a redesign of tube-side passes.
- **b.** Furnish new bundles using a different fin material, height, and spacing.
- **c.** Increase the number of plates in plate exchangers.
- **d.** Add double-pipe units in series or parallel or change their elements.

3. Often, alternate operating conditions govern, requiring a *new rating* to confirm the equipment selected meets both sets of conditions.

4. A time-consuming task is *determining why an exchanger is not performing*. Often the exchanger design and construction is the cause, not the surface furnished. The rating engineer should verify that the construction will be adequate before finalizing the recommendation. It usually takes more time to determine why a unit is not performing than is needed to rate 20 or 30 exchangers. This construction review is time well spent as it often avoids field problems. Further, non-performance adds the expense of field trips to the project. Several examples are cited later in this chapter and in Chapter 19.

5. *Alternative selections* may be needed for the choices described in items 1 and 2. The first choices may be too large to ship economically or may present handling problems, or they may not fit in the available space or not be easily serviced. An exchanger may require more water than is available, or existing piping may not handle the quantity needed. In these cases another selection has to be made.

6. The engineer should determine that *the selection is economical*. Otherwise, the selection is moot.

7. Often engineers must *choose a fix that will have a plant back onstream* rapidly even if it is not the best option.

Air-conditioning (A/C) systems present similar challenges except that exchanger sizes are usually smaller. Some selections, particularly evaporators, present condensation problems. Freezing is not an uncommon condition. In recent years evaporator and condenser choices have been influenced by the quantity of refrigerant in tubes because the refrigerant cost has increased. The quantity needed can be minimized using smaller-diameter tubes, which have less volume (Section 2.4.3), resulting in reduced refrigerant and system cost, though not necessarily exchanger cost.

Cooling of electronics differs because heat exchangers are not always needed. Typically, only the main A/C system exchangers are needed. Sometimes small A/C systems are installed in racks to provide local cooling. The cooling air directed across the equipment cools the electronics. Starving fans (Sections 12.4, 19.2-19.4, 19.11, and 20.2) and recirculation (Chapter 20) occur as frequently in this industry as in process or A/C work.

This book centers on making effective selections in terms of surface, good fluid distribution, operation, maintenance, construction, shipping, handling, and cost. If problems are to be minimized it is essential that information flow freely between buyer and seller. In the past minimum efforts have been made to train personnel, other than the rating engineer, in skills needed to solve problems and improve designs. It is the intent of this effort to pass along the skills learned in one generation to the next.

1.2 PERSONNEL INVOLVED IN IDENTIFYING REQUIREMENTS

The talents of many people are needed if exchangers are to operate successfully. Most plant personnel do not have the skill for selecting equipment but they have the talents to assist in successful installations. The needed effort is in making use of these talents. Many employees can assist by obtaining information. The examples that follow in this chapter show why it is necessary that requirements and limiting conditions be known before the selection process begins. An unstated requirement can have consequential effects that increase with exchanger size. Installing exchangers in tight spaces can be a problem. The cost of modifying structures or walls and relocating equipment can be high; at one installation it was more than five times the cost of the exchanger.

Many personnel contribute to successful installations. Sometimes individuals perform more than one service. Personnel involved frequently have these titles:

1. Specification writer
2. Heat exchanger buyer
3. Heat exchanger seller or the seller's representative
4. Rating engineer
5. Pricing engineer

6. Design engineer
7. Drawing checker
8. Shop engineer
9. Shop inspector
10. Handling, boxing, and shipping engineer
11. Buyer's inspector
12. Code inspector (ASME or other)
13. User's engineer or field personnel
14. Personnel who perform studies

A time-proven way of having a successful project is to have complete information on what is needed. This chapter will emphasize and illustrate three approaches toward the end of defining requirements completely. The first approach is to examine the application and to see what sets it aside from similar applications. Table 1-1 lists the likely requirements of five categories of air coolers to illustrate how they can differ. The purpose is to show that specifications written for one application do not necessarily apply to similarly designated units. In the second approach, personnel might ask buyers and users a series of questions on several subjects so that needs are not overlooked. The third approach is to look at the history of requirements that were omitted on other projects and the difficulties that their omission caused. In all three instances the intent is to furnish ideas for improving selections. An omitted requirement can result in the users' purchasing equipment that is not to their liking and does not meet their needs. Buyers and sellers want to avoid this experience.

The remaining tables in this chapter illustrate the second and third approaches to defining requirements. There are overlaps in the subjects noted but the basic ideas apply to all. Six tables relate to the interactions between the seller and the buyer or user, or the second approach toward writing the requirements. Personnel apt to have contact with the buyer or buyer's organization can use these to assist in obtaining information that better defines the requirements for a project. The information obtained may be sufficient to establish a competitive advantage or avoid a poor mechanical selection and costly error.

Table 1-2, Subjects Buyers and Sellers Should Agree On
Table 1-3, Questions Sales Personnel Might Ask Users

There is a significant difference in the operation of process and air conditioning exchangers. The former usually run continuously; the latter do not. Air-conditioned space frequently is subjected to different operating conditions than those for which they were designed. Examples are using the space for other purposes or a new tenant with different cooling requirements. These generally will not be known at the time the system is installed. Hence, requirements in A/C applications are different from those in process industries and these need resolution between buyer and seller. Some examples are given in the following table.

Table 1-1 Potential Requirements for Various Air Coolers or Heaters

P = potential, AHU = air handling unit, TBD = to be determined

Application	Compressor intercooler/ aftercooler	A/C evaporator	A/C condenser	Process air cooler	Motor cooler
Air distribution	X			X	
Alternate operating condition				X	
Aspect ratio		X	X	X	
Closed air system	X	X			X
Closure	X	X		X	
Condensation, outside tubes	X	X		X	
Condenser flooding (controlled)			X—inside		
Cost—supplemental		Refrigerant		Handling and shipping	
Drainage: Inside tubes	X	AHU possible		X	X
Outside tubes	X	X	X	X	X
Distributor		X			
Flashing			X		
Fluid in tubes					
Refrigerant		X	X		
Chilled water	X	X			X
Heavy HC				Channelling	

Freezing: Inside tubes	X	X—water			X
Outside tubes	X	X	X		X
Handling	X	X	X	X	X
Heaters (supplemental)		X		X	
Instrument couplings or vents	X	AHU—X		X	X
Louvers		By others	X	X	
Materials (special)				X	
Noise	P	X		X	P
Nozzle location(s)	X			X	
Operation: Continuous	X, TBD	X	X	X	X
Intermittent	X	X	X		X
Recirculation		X	X	X	
Refrigerant migration		X	X		
Scheduling	X	X	X	X	X
Shipping	X	X	X	TBD	X
Space limitations	X	TBD	TBD	Often	X
Transition pieces		X			
Starving fans or blower		X	X	Often	
Superheating		X			
Surface: Plate fin	Probable	Probable	X		X
Choice of spiral fin	X			First choice	X
Vibration	X			X	
Volume (internal)		X	X		

Table 1-2 Subjects Buyers and Sellers Should Agree On

(See also Table 1-9)
Missing requirements are often a cause of trouble. The following is a partial list of subjects frequently overlooked.Access for lifting equipment

Access to repair
Alternate operating condition(s)
Aluminum or copper fins
Auxiliary equipment not be part of exchanger)

- Manifolds
- Manifold supports
- Fans or blowers and their speed
- Louvers
- V-belt drives vs. gears
- Screen rooms (by others)
  - Floor height
  - Lack of insulation (under floor)
  - Overhead assembly space
  - Distribution (for two or more rooms)
- Boiling (unwanted)
- Bypassing
- Ceiling condensation
- Channeling
- Cleaning
  - Cover weight (light weight for ease of cleaning)
  - Fouling
  - Plate exchangers
  - Shell and tube sides
  - Pitch chosen
  - Tube length (effect of)

Cooling water temperature (or change in)
Cost
Cranes, lifts, hoists
Dehumidification
Delivery
Distribution

- Freezing or solidification
- Header size(s)
- Manifolds
- Unequal numbers of tubes per pass

Drainage
Drawings
Double-pipe exchangers
Energy usage/flow area
Evaporator

- Design
- Consequences if undersized

Exchanger: choice of air-cooler, shell-and-tube, double-pipe, or plate type
Face velocity
Fan coverage
Fans (banks of)
Filtration
Fins (external, internal, material)
Freezing
Fuel oil tank venting
Future expansion
Glycol solution(s)
Gradients (thermal)
Handling
Horizontal or vertical units
Hot spots
Hydrogen gas
Icing (see freezing)
Lift equipment and access
Limits (define various)
Log mean temperature difference (LMTD)
Lowfin tubes
Materials of construction
Mobile unit (exchanger suitable for application?)
Multiple exchangers to dissipate heat
Nameplates
Noise
Plate exchanger in place of chiller
Quality of construction
Radiation
Rating
Recirculation (water or air)
Recirculation chamber
Refrigerant (storage volume)
Relative humidity
Repair and access
Selecting type of surface
Shipping
Solidification
Space considerations
Spare parts
Split headers
Start-up and shutdown conditions
Storage tank (size, location)
Surface
Temperature drop across header (start-up, shut-down, operating)
Temperature extremes units will experience
Thermal gradient at start-up
Tubing temperature limits
Turbulators (use of)
Turnaround time (requirement?)
V-belts in induced-draft designs
Velocity

- Water
- Air

Venting

- General
- Fuel oil tank

Table 1-3 Questions Sales Personnel Might Ask Users
(See also Table 1-9)

Subject	Questions
Code vs. noncode unit	Are noncode units acceptable?
Drainage	Are drain couplings required? Can the outlet nozzle be used for this purpose?
Energy	Obtain the cost of energy. An efficient exchanger will be selected if the evaluation is on this basis. Otherwise, the low-cost exchanger is likely to be chosen.
Engineering	Identify where the engineering is done and the person to contact and phone number.
Expansion	Should the design include future expansion provisions?
Field or shop assembly	Will there be sufficient trained personnel to assemble large units at the site?
Freezing	Is freeze protection required? Air side? Water side? Product side?
Lift capacity	Determine limits. Exchangers may be selected to reduce the need for large cranes.
Metallurgy	Obtain a list of the preferred materials and unacceptable materials.
Refrigerant	Will refrigerant savings be included in the evaluation? (Section 15.1)
Schedule and downtime	Discuss client's needs. Usually they can be met if known in advance.
Shipping	Define method of shipping. Also, date needed at site.
Solidification	Is product solidification possible? Is preventive protection required?
Space	Define installation space and access route.
Spares	Determine type and quantity of spares required.
Unusual conditions	Examples: flaking (Section 15.2); tolerances (Sections 2.1, 2.4, 4.6, and 17.4; Tables 1.9 and 2.4); weight (Section 21.6).
Vents	Are vents required? Can venting be through inlet or outlet nozzles?
Warm air out	Confirm that warm air out will not be a hazard to personnel or affect nearby equipment.

Table 1-4, Features A/C Designers Should Consider

There are aspects to workable exchangers not covered in the specifications. It is essential these too be met if units are to perform. The previous tables refer exclusively to requirements; the following provide guidance for other personnel to assure that the units delivered will perform as guaranteed. Note that other personnel are instrumental in providing this assurance. The aids for these personnel are:

Table 1-5, Details Air Cooler Inspectors Should Look For
Table 1-6, Details Shell-and-Tube Inspectors Should Look For

Table 1-4 Features A/C System Designers Should Consider

Subject	Comment
Access to repair	Specify need and allow space.
Increasing system capacity	Consider leaving the existing evaporator as is, and add a makeup air cooler. This can be installed at minimum cost and without downtime (Section 16.3). Address the freeze problem it may introduce.
Is no outage time a requirement?	Is a spare A/C unit required? In some designs repairs can be made while the unit is running.
Outage time	Design so that modifications can be made while the units are off, as in winter, during evening hours, on weekends, and during plant shutdowns.
Positioning electronic equipment in racks	Have the heat exchanger engineer review rack layouts and flow areas before cables are manufactured to minimize future cooling problems.
Transferring A/C cooling from one room to another	The cost of this requirement is small, but correction later will be costly. Designers should include the need in the specifications.

Table 1-7, Details Checkers Should Look For

Two more tables illustrate the third approach, studying the record of past omissions that resulted in trouble jobs. Many of these are detailed in this book. The tables have the added benefit that they can be used by specification writers as a reference to check that no requirement is overlooked.

Table 1-8 Some Causes of Poor Thermal Designs or Selections
Table 1-9 Some Causes Of Process Units' Not Meeting Requirements

The tables suggest information to seek. Every subject listed was a trouble job at least once in the writer's career. Often a requirement that should have been written was missing when the selection process began. The goal is to identify missing requirements before their omission causes a problem. The tables are meant to provide ways of identifying requirements and hints of what to look for when units do not perform.

Contacts between seller and user are emphasized so that needs are not overlooked. A missing requirement can lead to a wrong selection or result in a costly correction. Several examples stress this point. While numerous selections can be thermally correct, one is usually best in terms of maintenance, construction, handling, manifolding, shipping, cost, or other criterion.

The responsibilities of personnel involved in the purchase, selection, installation and operation of exchangers differ, but all can contribute to near trouble-free installations. Manufacturers usually offer competitive selections. Users may reject an offering because it can't be cleaned mechanically, yet they themselves may have failed to identify this need as a requirement. How can an engineer know something is needed unless it is in the specifications, one may wonder. A common complaint of engineers is they seldom see

Table 1-5 Details Air-Cooler Inspectors Should Look For

Subject	Comment
Couplings	Confirm pressure and temperature gauge couplings are installed, if required. When two or more inlet or outlet nozzles are required, usually one inlet and one outlet require these. If couplings are plugged, make certain they are full depth to avoid thread damage.
Cover plates	Confirm cover plates have not been manufactured using end mills. This manufacturing method tends to create grooves that become leakage paths.
Drainage	Passes should be designed for drainage to avoid product freezing, solidification, or buildup or other obstructions. Weep hole(s) may be needed.
Fin bonds and ends	Confirm that the fin bond is adequate and does not lead to fin unraveling.
Galvanizing	Section 4.1 and Chapter 13.
Lifting lugs	Lugs shall be located so that sections can be lifted without buckling. Is a spreader bar required? Will it be available when needed?
Match marking	Field assembly is simplified when sections and components are properly marked. Is the equipment packaged to avoid damage during shipment? This applies to fans, fan blades, panels, fan rings, walkways, and, particularly, sections requiring manifolds.
Nozzle faces	Confirm nozzle faces have not been scraped, scratched, or nicked and are protected during shipment. Do bolt holes straddle the centerlines, and are nozzle faces parallel to the ground? When headers are split for drainage, nozzle faces shall be parallel to the ground and not the header box.
Oversized plugs	Damaged plug holes are often redrilled and fitted with an oversized plug. Confirm these are properly identified.
Paint	See under side channels.
Pass partition plates	Do these have continuous welds at the end plates to avoid bypassing?
Seal strips	Are these installed (Section 19.13 and Figure 19-7)?
Section geometry	Confirm the section is rectangular and not warped in a rhomboid shape.
Shipping pins	Shipping pins are designed to keep the floating header in position during shipment. Are they marked for removal once the section is installed?
Side channels	Confirm these are painted or galvanized before assembly as this usually cannot be done after assembly. Structures must have continuous welding for uniform coverage during galvanizing and must be sturdy to prevent warpage during this process.
Tolerances	The cumulative width of sections placed side by side should not be greater than the pipe rack space available. Check tolerances (Section 2.1).
Tube supports	Confirm these are aligned and hold position.
Vents	Confirm these have been provided.
Weight	Check the weight calculation. If minimum wall tubes have been installed, make sure average wall weights were not used in calculating weight.
X-rays	Inspectors should decide where X-rays are taken.

Table 1-6 Details Shell-and-Tube Inspectors Should Look For
(See also Section 19.13.)

Subject	Comments
Baffle tolerances	Section 19.14.
Bundles	Bundles should slide in and out easily. Clearances between shell ID and baffle OD should be minimal (Section 19.14).
Bypassing	Section 19.13.
Channel cover or floating head cover	Have threaded holes to lift heavy covers been provided?
Components	Have the following been provided where required? Couplings for instruments Drains and drain plugs Tube supports or baffles Impingement plate Vents
Corrosion protection	Surfaces to be painted should be sandblasted and given a protective coating unless painted immediately. Has this been done?
Couplings	Confirm that pressure and temperature couplings are installed if required. Full-depth plugs should be provided to avoid thread wear and damage.
Drainage	Passes should be designed to drain fluid and avoid freezing, solidification, product buildup, or other obstructions. Have weep holes been provided, if needed?
Jackscrews	Have these been provided (if needed) for separating large flange connections?
Mounting pads	Mounting pads shall not be too close to the channel and shell inlet nozzles. Tight tolerances may result in cocked flange faces, causing leakage. This is particularly true when ring-type joints are used (Section 17.4).
Nozzle faces	Nozzle faces shall not be scratched, scraped, or nicked, and shall be protected for shipment. Bolt holes shall straddle centerlines. Confirm that flange faces will mate with connecting flanges where applicable.
Pass partitions	Confirm pass partitions are continuously seal-welded to prevent bypassing.
Pins and gaskets	Pins shall be provided, where necessary, to hold gaskets in position during assembly.
Seal strips	Have seal strips been installed? They are required in most cases.
Stacked units	Units shall be shop-assembled and match-marked before shipment.
Vents	Confirm vents have been provided where needed
Weight calculation	Confirm weight calculations. Make sure that minimum wall tube weights are used when these are furnished. For steel, minimum wall tubes weigh about 10% more than average wall tubes.
Welds	Welds shall be inspected for full penetration or for other required construction.
X-rays	Inspectors shall specify where X-rays are taken

Table 1-7 Details Checkers Should Look For

Subject	Comment
Construction	Confirm construction required is furnished (ASME, API-660, API-661, TEMA, etc.).
Couplings	Have these been furnished on inlet and outlet nozzles (for pressure and temperature gauges) if required?
Footings	Confirm footings are adequate for current and future requirements. A future add-on could double the load on a footing (Section 21.9).
Headers	Confirm header thicknesses are adequate for the design pressure and temperature. Question thin parts.
Maintenance	Is space provided to repair, maintain, install, or clean equipment?
Manifolding	Size: Adequate? Supports: Are they required? Distribution: Adequate, or adaptable to expansion?
Noise limits	Confirm that noise level guarantees meet specifications.

Operating conditions	Many potential problems can exist. Here are some of them:
	• Start-up
	  • Refrigerant migration
	  • Thermal expansion
	• Shutdown: Have these been addressed?
	  • A freeze situation
	  • A solidification situation
	  • A refrigeration migration problem
	  • The need for control in cold weather

Recirculation	Verify that, as installed, recirculation should not be a problem.
Seal strips	Are these installed (Sections 19.13 and 19.14)?
Space	• Do cumulative tolerances create a problem? • Are manifold supports a requirement? • Does a potential recirculation problem exist? • Is there space for bundle removal? • Is sufficient space available for access needs (cranes, hoists, cleaning equipment, removal, expansion, etc.)?
Tube diameter	Question the use of large-diameter tubes in refrigerant service. These add substantially to the quantity of refrigerant needed to operate a system. See Section 2.4.6.
Tube material	Confirm the water velocity in the tubes used.

their recommendation in service because selections are made at locations removed from plant or installation site. When they see their selection in service it is surprising how often they say, "We would not have chosen this design had we known this limitation" (unspecified need). To avoid this kind of slipup, Table 1-2 contains pointers to help in identifying requirements. Needed information can be obtained from the buyer or from site, sales, or marketing personnel.

Buyers and sellers want to avoid trouble. If enough questions are asked, the requirements will surface. The tables developed are not all-inclusive.

Table 1-8 Some Causes of Poor Thermal Designs or Selections

Access or repair space not provided
Alternate operating condition not considered
ASME or non-ASME construction not defined
Cleaning, construction does not allow for
Coils of the same external dimensions but not the same tubes and fins may not perform equally
Cooling or heating load changes
Cost
Counterflow rather than parallel flow to minimize freezing or channeling
Crane rental, eliminate need for
Delivery time too long
Downtime, potential for not minimized
Drains not provided
Energy, uses too much
Evaporator undersized
Fan
- Coverage
- Recirculation
- Starving

Fin choice
Flow restrictions not fully considered
Forced or induced draft, advantages not maximized
Fouling factors, choice of
Freezing
Headers undersized
Lifting lugs not provided
Maintenance, excessive
Makeup air, insufficient amount
Manifolding not well chosen
Material, poor choice of
Mating equipment limits met
Mixing, adequate, not provided
Noise, equipment sized without considering
Overcapacity at some operating conditions
Parts availability
Plant expansion not provided for
Pressure drop, cumulative effect
Radiation, thermal, not considered
Recirculation
Refrigerant, quantity of
Requirements, not all included
Shipping problems not addressed
- Heavy loads
- Shipping plugs to minimize shock loads
- Wide loads

Space considerations not met
Spare parts not adequately defined
Stacking units and seal integrity
Supports too close to nozzles
Temperature drop too large in one exchanger
Tolerances not considered
Tubing material selected not the best choice
Velocity limits
- Air
- Water

Vibration

Every project has needs that should be in the specifications (provided they are in the contract, as requirements cost money) if the best selection is to be made.

1.3 REQUIREMENTS BY APPLICATION

As already noted, one way to completely identify requirements is to pay close attention to the application. Table 1-1 listed requirements of five different air cooler applications. Each had its own different requirements. The lesson is, Do not assume that specifications written for one application apply to another.

Table 1-9 Some Causes of Process Units Not Meeting Performance Needs

- Blockage
  - Flaking (Section 15.2)
  - Mud (Section 19.7)
- Blockage, lack of
  - Welds omitted (Section 19.13)
- Bypassing (Section 19.13)
- Channeling (Sections 2.3, 12.8, and 19.15)
- Cleaning (see also Fouling)
  - Brushes (Section 14.5)
  - Effect of tube length and mechanical cleaning on cost (Section 2.2)
  - Furfural cooler (Section 15.3)
- Clearances
  - Air coolers (Section 2.1)
  - Baffles (Section 19.14)
  - Section widths (Section 2.1)
  - Shell and bundle (Section 19.14)
  - Shell, assembly of (Section 17.4)
- Condensation (Sections 18.2 and 26)
  - See also Freezing (Chapter 12)
- Cost too high
  - Changing metallurgy (Section 15.9)
  - Tubing (Section 15.5)
  - Welded shell (Section 15.9)
- Data published in error (properties) (Section 19.15)
- Design and operating conditions compared
  - Parked trailer operating condition overlooked (Chapter 22)
  - River water temperature (Section 19.15)
  - Turbulent to transition to laminar flow (Section 2.4)
- Drains (Section 12.2)
- Energy usage (Sections 15.8 and 17.5)
- Fan coverage (Section 6.1)
- Features that should make selections unacceptable (Chapter17)
- Flow, lack of obstructions to (Section 19.13)
- Flow restrictions
  - Flaking (Sections 1.4 and 15.2)
  - Header size for lube oil (Sections 11.2, 11.4, and 19.5)
  - Mud (Section 19.7)
  - Plate exchangers (Section 19.7)
  - Shell-and-tube exchangers (Section 2.4 and 15.2)
- Fluid properties
  - Using old data (Section 19.5)
- Fouling (Chapter 10)
- Freezing (Chapter 12)
- Header size (Sections 11.2 and 11.4)
- Lack of obstruction to flow (see Flow, lack of obstruction to)
- Laminar to turbulent flow (Section 2.4)
- Manifolds and headers (Chapter 11)
- Mounting pads (Section 21.9)
- Noise (Sections 4.2 and 17.3)
- Obstruction to flow (Sections 19.2 through 19.12)
- Particle size (Sections 8.1, 15.2, and 19.6)
- Piping (see Manifolds)
- Pressure drop
  - Cumulative (Sections 19.8 thru 19.10)
  - Distribution (Section 19.5)
  - Inlet nozzle location(s) (Sections 11.2, 11.4, and 11.5)
- Recirculation (Chapter 20)
- Requirements
  - Missing (Sections 1.3 and 1.4)
  - Defining what is not wanted (Section 1.5)
- River water temperature (Section 19.15)
- Seal strips (Section 19.13)
- Shipping pins or plugs (Section 21.1)
- Shock load (Section 21.1)
- Stacking units (Section 17.4)
- Starving fans (Section 20.2)
- Surface, adding to minimize downtime (Chapter 16)
- Tank heater installation (Section 7.2)
- Tolerances (Sections 2.1 and 17.4)
  - Baffles (Section 19.14)
  - Section width (Section 2.1)
  - Shells (Section 19.14)
  - Stacking units (Section 17.4)
- Tube-wall viscosity (Section 18.1)
- Water velocity (Section 17.2)
- Welds omitted or poor seals
  - Pass partition plates (Section 19.13)
  - Two-pass shells (Section 19.14)

1.4 PROBLEMS DUE TO A MISSING REQUIREMENT

In Chapter 19, "Thermally Correct Systems That Do Not Perform," several installations are cited. Many comments in that chapter apply here. The following examples are of other units that did not perform because mating equipment requirements were not met or because one or more requirements were missing. Poor economic choices are included because there is a relationship between engineering and cost. The examples are from other chapters of the book and are assembled here to illustrate the point. Thermally correct selections are assumed.

The number of units that do not perform is small compared with those that do. Sufficient data are usually provided to make a good selection. Requirements cost money. Sellers driven to obtain orders usually offer their lowest-cost selection. They are aware that adding one requirement may make their offering noncompetitive. They could provide an alternate recommendation (costly to seller), but in the final analysis requirements are the buyer's responsibility. Exchangers should be looked at in perspective. Neglecting mass-produced units, which nearly always perform, only about 1 in 10 selections need some correction but 1 in 25 present a serious problem that cannot be overlooked. Rarely is the cause poor sizing.

Serious problems can occur in existing installations. Here the concern is not sizing of equipment but performance. It takes time to identify the cause of a problem, and time is usually not available. Making the correction rapidly is the goal. The problems that follow resulted from omitting a requirement or not completely defining the need. Problems are usually not thermal in nature unless they relate to an alternate operating condition. The manufacturer is seldom aware that a requirement is missing unless informed in some manner. A new selection can usually be made to include a requirement. The following cases represent examples where a requirement omission caused a problem.

1.4.1 Furfural Cooler (See Section 15.3)

In an instance to be described in Chapter 15, the specifications should have included cleaning, its frequency, cover weight, maintenance schedule, and assembly and disassembly. Because these were not given the correction was to scrap the exchanger and buy a new one.

Conclusion: Cleaning needs should be specified.

1.4.2 Quench Oil Cooler(s) (See Section 15.2)

As an inevitable result of the industrial process in a metal stamping plant, large metal flakes were present in the quench oil, but these were not noted in the specifications. The flakes blocked flow passages, preventing the exchanger from performing. The correction was to purchase new exchangers using air rather than water for cooling.

Conclusion: The presence of the particles should have been in the specifications.

1.4.3 Manifold Supports (See Section 2.1 and Chapter 11)

In one application, the need for extra structural members to support a manifold was not given. Such additional supports may take away space intended for the heat exchanger. This requirement will almost always impose constraints on the selection and require structural changes.

Conclusion: Structural needs belong in the specifications.

1.4.4 An Underperforming Environmental Control System (See Sections 19.3 and 19.4)

It is reassuring to read an environmental control system (ECS) nameplate and know the unit was designed and guaranteed to deliver the cubic feet per minute and the cooling needed. In the cases discussed in Sections 19.3 and 19.4, the units delivered did not meet the nameplate conditions. An ECS's performance also depends on mating equipment. When a unit does not perform, the engineer is asked why. Frequently, undersized ducting is the cause. Here is an example. An ECS blower is to deliver 10,000 cfm at 1.0 in pressure loss (0.5 in of this is taken in the ECS evaporator and filter; the other 0.5 in is for the ducting). The ECS guarantee is 10,000 cfm at 0.5 in (1 in – 0.5 in). The duct pressure loss is 1.1 in, and the evaporator and filter loss is 0.5 in. Thus the pressure the blower must supply is 1.6 in, not the 1 in it was designed to deliver. At the installed condition with more duct resistance, a blower delivers fewer cfm, perhaps fewer than needed. No exchanger requirement is missing. The allowable duct system pressure drop is exceeded and the blower cannot deliver the required airflow. Hence, the ECS will not meet performance (see Figure 1-1).

In most cases ducting is sized correctly but unexpected field problems occur requiring that changes be made. Most cause an increase in line pressure loss because space is not available to make the best modification. This loss can be large. At one installation the blower's cfm output was only 41% of design because both ducting and evaporator were undersized. Sound-deadening material, added later, further reduced flow. In trying to determine why an A/C or ECS unit is not performing, a check of system pressure drop is a good place to start. Be aware this is a time-consuming effort (Sections 19.3 and 19.4).

Conclusion: An undersized duct system can be the cause of underperformance in ECS or A/C units.

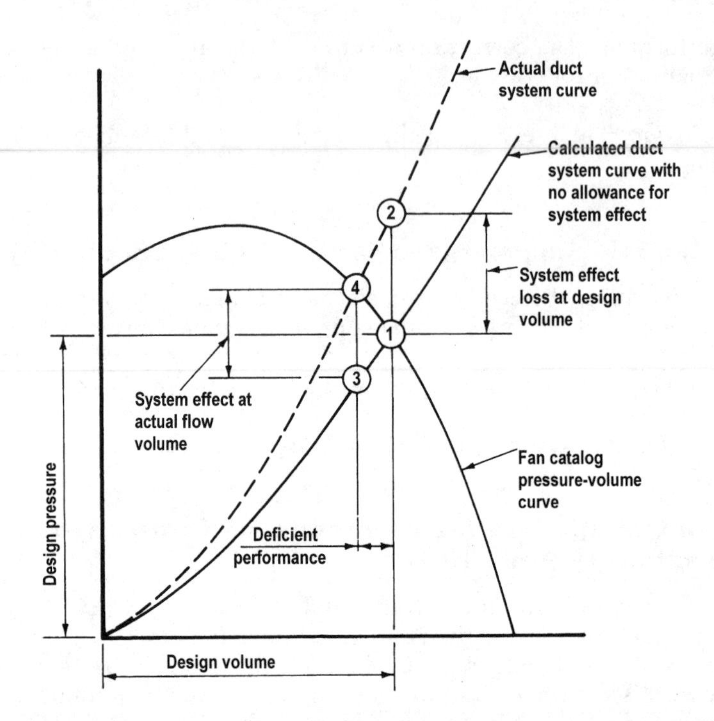

Figure 1-1. Deficient fan/duct system performance. System effect ignored. (Reprinted from AMCA 201-90, *Fans & Systems,* with written permission from Air Movement and Control Association International, Inc.)

1.4.5 Future Expansion (See Chapter 16)

The time to address a future expansion if these details are known is when the initial selection is made. One choice for increasing a plant size by 50% is to use two shell-and-tube units in parallel and add a third later. If the initial selection (not considering expansion) has one or three units instead of two, no symmetrical manifold selection will satisfy this mix. Planning has benefits.

Conclusion: Include future requirements in the specifications if known.

1.4.6 Shipping Weight (See Section 21.6)

Two smaller but lighter units instead of one large unit may eliminate the need for renting a crane. A plant's lift capacity should be in the specifications so that selections can be made to eliminate a crane rental cost if possible.

1.4.7 Shipping Width (See Section 21.5)

Limiting dimensions on-site or along the access route should be in the specifications. One defines the exchanger's limits; the other, its handling. Handling charges have, at times, exceeded the cost of the exchanger.

Conclusion: Including this kind of data in the specifications is a potential way of reducing cost.

1.4.8 Delivery (See Section 15.7)

Shipping, handling, and delivery limitations often present problems difficult to solve. Such data should be in the specifications. Some designs are available in shorter times than others or are more easily managed in terms of shipping and handling.

1.4.9 Shock Load and Surging (See Section 21.1)

Shock loads and surging are common in applications designed to remove the heat of compression from gases. Plate exchangers are not suited for this service because of their thin plates. Compressed gases should not be placed on the shell side of shell-and-tube exchangers unless the baffles are designed and spaced to avoid the critical harmonic frequencies of the tubes. The requirement is, The exchanger shall be capable of sustaining the shock loads and surges associated with compression.

Conclusion: Define all operating conditions.

1.4.10 Noise (See Section 17.3)

Noise is an ongoing problem best solved by including its limiting conditions in the specifications.

1.4.11 Economic Competitiveness

A selection must be competitive. Exceptions occur such as a need for rapid delivery, light weight, adapting the unit to the available space, or providing large corrosion allowances, to name a few. An understood requirement is, Select the unit with the lowest cost.

Conclusion: All requirements should be in the specifications. These vary with the application or service.

1.4.12 Turbulators (See Section 14.4)

The heat transfer rate of fluids in laminar flow or in the transition region can be improved using turbulators. Turbulators are not suitable in applications where significant fouling exists unless they can be chemically cleaned. This applies to interrupted fins as well (Section 7.4). If fouling constraints are not expressed, turbulators are apt to be offered because they can reduce the cost of an item. Their use may create cleaning and maintenance problems.

Conclusion: Define the attendant fouling in the specifications.

1.4.13 V-Belts and Partial Load Conditions

In some induced-draft coolers, fans and V-belt drives are mounted on a fan ring. This design is not recommended in certain applications. Assume a product is to be cooled from 300°F to 150°F. At partial load, the exchanger will be oversized, resulting in reduced airflow. The outlet air temperature can rise to over 200°F, placing the V-belts in a high-temperature atmosphere and thus shortening belt life. In this application the forced-draft unit is preferred. Both units are selected the same way. Experience teaches that one design is preferred over the other in some applications.

Conclusion: Define alternate operating conditions.

1.4.14 Critical Pressure Drop (See Sections 19.3, 19.5, 19.9, and 19.10)

Examples of critical pressure drop situations are given in Chapter 19. The sections are titled "Screen Room Ductwork," "Chiller Plant," and "Refinery." Each illustrates a major failure.

Exceeding an exchanger's allowable pressure drop may not be a problem. When it is a problem it is costly to correct and usually results in a reduced plant output. The allowable pressure loss over several exchangers in series should not be exceeded.

Conclusion: Define the allowables and confirm these are met.

1.4.15 Storage Tank Exchangers (See Section 7.2)

Before purchasing a storage tank exchanger, confirm that draining the tank to remove a bundle is an acceptable operating condition. The question must be answered, Where will the product be stored so that the exchanger can be removed? Is a spare tank available? To minimize the need to drain a tank, bare coils are made of heavy Sch 80 pipe (Figure 7-2). Thick pipes provide a factor of safety to keep tanks on-line. If the removal of a bundle is expected, have the shell cover fitted with a nozzle and shutoff valve so that the tank need not be drained to service a bundle.

Conclusion: Include site-specific requirements in the specifications.

1.4.16 Refrigerant Storage (See Section 15.1)

An exchanger with small-diameter tubes can reduce the quantity of refrigerant needed to operate a system, thus minimizing system cost. The selection may not provide the most economical exchanger but often results in the most economical system, a feature to be noted when proposing an exchanger.

Conclusion: Define the minimum acceptable tube diameter, and inform personnel of its economic benefits.

1.5 REQUIREMENTS THAT DEFINE WHAT IS AND WHAT IS NOT WANTED

Specifications should include what is and what is not wanted. Here are examples of statements whose absence has caused trouble:

1. The exchanger shall not contain copper or copper-bearing parts. (When this wording is used it should be clear if the exclusion of copper also applies to electrical equipment such as motors or circuit breakers.)
2. The noise level shall not exceed 80 dBA at 5 feet in any band. Personnel should not have to work in high-noise environments or wear earplugs.
3. Auxiliary heat shall not be furnished. (This could be a deduct item reducing the cost of a standard unit.)
4. V-belt drives shall not be used over 30 horsepower.
5. Tube lengths for air coolers shall be 32 feet. Shell-and-tube unit tube lengths shall be 20 feet. This is for consistency with other plant exchangers.

1.6 REQUIREMENTS FOR FLUID PROPERTIES

Fluid properties and conditions should be given. Usually the physical and chemical properties of most fluids are known and readily available. It is possible that even without a complete set of these the selected exchanger will meet performance, require infrequent cleaning, and be easy to maintain. If, on the other hand, particles of undefined sizes are present, the odds of facing a trouble job are high. The engineer should be aware of the presence of these particles to avoid problems. Examples of how their presence affected the operation of some exchangers follow.

The quench oil cooler (Sections 15.2 and 19.6) did not perform because the presence of large flakes blocked flow passages. In another example when repairs were made to a water supply system large clumps of mud entered the pipe and flowed to a plate exchanger where these blocked flow to surface causing the unit to underperform (Sections 8.3 and 19.7). In some refinery applications fine particles flowing in the fluid cleans surface, somewhat like sand blasting, minimizing the need for large fouling factors. There are refinery applications where high velocity particles in the flowing stream cause pitting particularly at the entrance to exchanger tubes. Users should advise manufacturers of their experiences with flowing particles to avoid repeating this kind of problem. The nature and chemistry of a fluid must be known if brush cleaners are to be used effectively (Section 14.5).

1.7 SUMMARY OF PROBLEM AREAS

As emphasized repeatedly in this book, a selection may be thermally correct although the construction chosen is not the best choice. There are many questions that do not have to be answered just to size an exchanger but whose answers are necessary to make good recommendations. Their frequent omission from specifications has resulted in problems such as underperformance, excessive cost, or difficulties in shipping and handling.

A good checklist of often unspecified requirements would include the following:

Access to site
Air filters
Air velocity limit
Air vs. water cooling
Bundle removal space
Channeling
Cleaning lanes in shell-and-tube exchangers
Cost
Cleaning frequency
Code stamp
Cranes needed to install
Cumulative pressure drop
Delivery
Distribution
Dome area
Duct design
Energy
Extended surface
Fan coverage
Field/shop assembly
Flange faces (type)
Fouling
Freezing
Impingement plate
Installation
Insulation
Maintenance
Manifold supports
Manifolding
Metallurgy
Missing a requirement
Mounting pads
Noise
Particle size
Potential for blocking flow
Provision for expansion
Radiant energy
Reduced-flow operating condition
Refrigerant storage
Relative humidity
Shipping
Shock load
Tampering with controls
Thermal gradients
Tolerances
Undersized evaporators
Vibration
Water velocity
Weight

The author has addressed all of the problems in this list, and most of them several times.

1.8 CLEANING AND FOULING

Consider fouling in these terms:

- Performance of the unit
- Frequency of cleaning
- Ease of maintenance
- Particle sizes present

If a unit does not perform to specification, it is a trouble job. Excessive cleaning frequency is trouble of another sort, and difficult maintenance of yet another. The ultimate trouble is anything that blocks access to the exchanger surface, preventing it from performing. Some problems are easily corrected, while others are difficult to determine. Corrections are costly, some more than the exchanger. The consequential loss of being off-line may be greater than either.

Consider an exchanger with a requirement that reads, "The tube side shall be capable of being mechanically cleaned." Some exchangers in food service are cleaned daily, while others run a year or more between cleanings. Yet, inattentive writers might come up with the same requirement for both cases though each has a different need. Hence, the requirement should be more specific. Covers that are removed frequently to have access to tubes should be light in weight and have simple gasketing (which, in turn, presupposes fewer pass partition plates), and it should not be necessary to disconnect inlet and outlet nozzles to remove these covers. Heavy covers, a large number of passes, and complicated gasketing add to cleaning time. Cleaning and handling should be simple. These specific needs are less significant, however, if servicing is done at intervals of a year or more. Thus, while requirements that read the same may be literally correct in both cases, more attention to the details may reveal that one design is preferable to another.

Thus, when frequent cleaning is required, include in the specifications that a lightweight cover is preferred. It is easier to remove a 75-pound cover than one weighing 300 pounds. Exchangers can be selected to minimize the weight of this part. On the other hand, it is less work to clean a unit once a year than once a month. Why not solve the problem of excess maintenance when the exchanger is selected? This can be done in two ways: add more surface to allow more fouling, or select a unit to simplify the cleaning effort. (An example is given in Section 15.3.)

The shoulder-plug air cooler design is often preferred for fluids that operate at relatively high pressure, particularly when frequent cleaning is required. Plugs are easily removed. It is a good idea to specify the desired construction if known.

1.9 CONTRACT REQUIREMENTS BETWEEN SELLER AND BUYER

The following suggestion is given to reduce tension between buyer and seller. Buyers are aware that funds for purchasing equipment are limited.

Frequently, buyer's and seller's engineers address a problem, agree on a fix, and proceed on this basis. This can be a seller's mistake. Always obtain the buyer's approval before making changes. Oral confirmations are not enough. It is surprising how often one of the parties to the agreement changes jobs, is transferred, is forgetful, or dies. No record of their agreement exists. The buyer may prefer the upgraded unit but may not have sufficient funds for this purpose. Without confirmation, the improved product may be delivered "free," resulting in trouble for the seller.

Chapter 2

Considerations before Finalizing the Selection

Once a heat exchanger has been selected, the rating engineer should ask if there is any reason why the equipment chosen is not a good choice for the application. In the first part of this chapter, two examples illustrate the point. Each cites conditions affecting the selection that often favor making another choice. Both cases emphasize that attention to detail minimizes the potential for costly errors. While adding a requirement may change the recommendation for a large exchanger, it is equally true of smaller units. These too must be lifted and maneuvered when space is limited. Space considerations are emphasized, but other factors affect a selection as well. In the final analysis, conditions that can limit a selection should be in the specifications.

2.1 CONSIDERATIONS BEFORE FINALIZING THE SELECTION

Once an exchanger has been selected and before it is finalized, consideration should be given to moving it to the site. Transport and installation are relatively easy when an installation is under construction, but they may not be so easy when a facility is being modified. At new sites, lifting equipment is on hand, and access is not as limited as it may be later. Some items to consider are:

- Is adequate lifting equipment available?
- Is the travel path wide enough?
- Is space available to make turns?
- Is space available for the unit selected and for bundle removal?
- Is space available to place lifting equipment in position?
- Will an extra shipping charge apply during transport from plant to site (for wide load, heavy weight, special handling, or flagmen, for example)?

The point is to identify problems before they occur. They may be minimized by making another selection. Here are some problems that occurred despite engineers' best efforts, though everything on paper was in order:

- *Tolerances (generally speaking, a manufacturer's problem)*. Suppose 96 ft of pipe rack is available for cooling sections. On it are placed 12 air cooler sections, each 8 ft 0 in wide. As built, the section widths averaged 8 ft 3/8 in. The pipe rack space needed is more than the space available by 12 times 3/8 in, or 4.5 in. The project engineer should have known this space-limiting requirement and could have avoided the problem by adding a note to the shop drawing stating that the section width, 8 ft, must be met with no plus tolerance allowable. This dimension is within the supplier's control. In the next example, the section spacing is not.

- *Manifold supports (generally, a buyer requirement that should be specified)*. A critical space problem on a similar application occurred when a large-diameter supply manifold was required. Consider the twelve 8-ft 0-in wide sections discussed in the previous example. The user determined, after the equipment was built, that the manifold and its 90° feeder lines were too heavy when filled with liquid to be supported on the headers alone. To support this manifold, columns that extend upward from the cooler structure are needed. These require a space 3 in wide between sections or pairs of sections for supporting steelwork. The columns reduce the amount of pipe rack space available. They require space that no longer exists for the exchanger purchased. Once a section is built, it is difficult to modify. The problem could have been avoided had the requirement been known before the equipment was built. Another selection with a smaller overall width would have provided space for these supports. The real difficulty is identifying requirements before equipment is built. The correction was to add a "doghouse" or overhang on the pipe rack structure, but in the final analysis this addition is costly.

Each of these problems could have been avoided had the requirement been in the specifications or had the project engineer been more alert. This is a challenge for specification writers and engineers. In cases where a requirement is not part of the "as built" equipment, the usual remedy is to modify the installation to make what has been built usable. This action will get the plant on-stream sooner but not without added cost.

2.2 EFFECT OF TUBE LENGTH AND MECHANICAL CLEANING UPON COST

A selection often begins with a determination of the maximum tube length that can be used, since the choice affects cost. Compare two shell-and-tube exchangers. Each has tubes of the same diameter, material, and thickness that are installed on the same pitch. One has four hundred 10-ft 0-in tubes, and the other two hundred 20-ft 0-in tubes. Neglecting tubesheet thickness, each has the same nominal surface. The shorter-length exchanger will have twice the number of tube-to-tubesheet connections and twice the number of baffle holes (though probably fewer baffles). Further, it will have a larger diameter and potentially thicker shell, shell flanges, baffles, channel, floating head cover, channel flanges, channel cover, and tubesheets. For these reasons, the shorter-tube exchanger with more connections and thicker parts is more expensive than one having longer tubes. Similar reasoning applies to air coolers, double-pipe exchangers, or other designs that use tubes.

If mechanical cleaning of tubes is required, space will be needed for cleaning rods and brushes. Compare these exchangers. The first, 10 ft 0 in, is installed in a room 24 ft 0 in long in the direction of tube length. The second, the 20-ft 0-in tube exchanger, when installed in the same room is near the maximum length that can be used, as space will be needed for a channel, channel cover, and shell cover as well as for walking around the unit. Should mechanical cleaning of tubes be required, space will not be available to place the cleaning rods and brushes inside the 20-ft tubes, while it will if 10-ft 0-in long tubes are chosen. Thus, the longer-tube exchanger would not be acceptable because it doesn't satisfy cleaning needs. However, the longer exchanger could be used if chemical cleaning is sufficient.

Other conditions may also allow the use of longer tubes:

- Space near the end of an installed exchanger used as a walk path could be used for maneuvering cleaning rods into position, thus allowing for somewhat longer tubes.
- Cleaning rods are sometimes made in two or more pieces with ends threaded. They are designed to mate with extensions. These shorten the cleaning rod length and effectively allow the use of longer tubes than if continuous-length cleaning rods are used.
- Power plants frequently use brushes, roughly 3 in long, to clean tubes. They are hydraulically propelled through the tubes while the plant is operating. The brushes are collected at the exchanger outlet. The process is repeated at intervals of a few hours. This procedure not only keeps tubes clean but reduces the tube-side fouling factor. In effect, these brushes improve the heat transfer rate and allow the use of an exchanger with less surface (see Section 14.5).

The necessity of cleaning the shell side as well as the tube side needs to be considered in the selection of tube length if the bundle is to be removable. The bundle length (tube length) and the shell from which it was removed fit in the available space in the present example provided that the shorter-length tube is chosen. It follows that cleaning the shell side is part of selecting the tube length, because space to clean it must be provided.

In refineries or chemical plants it is common to use the same tube diameter, length, gauge, and material in a facility because this reduces the number and variety of spares to be kept on hand. It permits more efficient storage while minimizing cost. Plants are designed so that tubes can be mechanically cleaned. The contractor will usually define the tube length and cleaning procedure.

Conclusion: The engineer should confirm what method, if any, of tube-side and shell-side cleaning is required before beginning the process of selecting an exchanger. This is part of the tube-length decision. Generally speaking, if an exchanger does not have to be disassembled for mechanical cleaning, the longer-tube-length exchanger is chosen because of its cost-saving potential. It should be understood that if sufficient space is available for cleaning rods, these factors do not apply and the least expensive exchanger can be selected.

2.3 PARALLEL FLOW

A parallel-flow exchanger is seldom recommended because it is the least efficient exchanger in terms of driving heat from one fluid to another. However, there are applications where its use has benefits, particularly where limiting the transfer of heat is an asset. The decision to use parallel flow should be made before the selection process begins. Some advantages follow:

1. The outlet temperature of the fluid being cooled can reach a limiting temperature if parallel flow is used. The advantages that follow are:
   a. If water is kept above 32°F, freezing will be avoided.
   b. Similar reasoning applies in preventing channeling (see Section 19.15).
2. Parallel flow can limit the temperature drop of air being cooled, thus keeping its moisture content high. This feature is useful in processing some commercial products and has been used to simulate tropical conditions for plants and animals in arboretums and zoos.

While this arrangement can be beneficial, there are conditions that reduce the limiting temperature that could create freeze or channeling problems. The most obvious occurs at shutdown.

Suppose water is cooled from 120°F to 60°F in an air cooler using air at 0°F as the coolant. The air outlet temperature for the design chosen will be 50°F. This condition is shown for parallel and counterflow in Figure 2-1. The counterflow exchanger has the following disadvantages (Figure 2-1*a*):

1. The unit is selected for the fouled condition but may be performing at clean conditions. In this case more heat will be transferred that will lower the water outlet temperature.
2. If the ambient temperature is reduced to –15°F, the counterflow exchanger will lower the water outlet temperature and increase the chance of freezing.
3. If water flow through the exchanger is reduced, a counterflow exchanger increases the chance of freezing.
4. At shutdown the exchanger will reach ambient or freeze temperatures sooner.

Now consider parallel flow for the same conditions (Figure 2-1*b*):

1. If the inlet air temperature drops to –15°F, more heat will be transferred and the air temperature through the exchanger will rise more than the current 50°F.
2. The terminal temperature difference was only 10°F and will drop as the quantity of heat transferred increases.
3. The mean temperature difference (MTD) for counterflow is 47% larger. This makes the needed parallel-flow exchanger larger.
4. Both parallel and counterflow units face freeze conditions following shutdown. The chance that this will occur increases with parallel flow because the exchanger is larger.

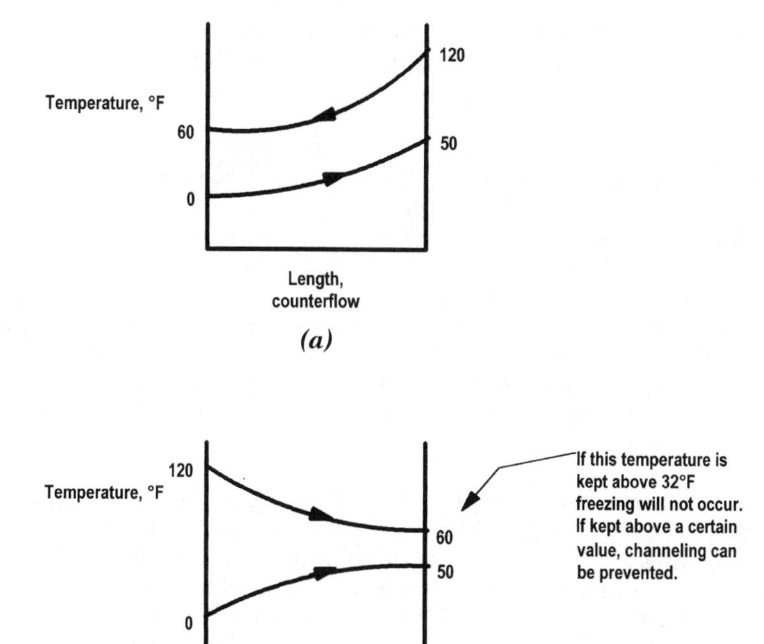

Figure 2-1. Axial temperature profile in a heat exchanger: (*a*) counterflow; (*b*) parallel flow.

Parallel flow minimizes the chance that freezing or channeling will occur but does not eliminate the possibility of either. Adding supplemental heat usually solves these problems (see Chapter 12, "Freezing").

2.4 TUBE-SIDE CONSIDERATIONS IN SELECTING AN EXCHANGER

An efficient heat exchanger is usually one that makes maximum use of the allowable pressure drop on both streams. Modifications to the tube side, assuming a fixed tube length, are limited to changes in the number of passes, in the tube diameter, or in the tube gauge. A beneficial variation, when permitted, is for exchangers in series to take advantage of the unused pressure drop allowed for one exchanger and apply it to other exchangers in the series. The increase can be apportioned in any way, provided that the allowable pressure drop through all exchangers is not exceeded. This leads to a more efficient bank of exchangers. The pressure loss apportioning applies to both streams. Caution: The total pressure loss allowed through all exchangers must not be exceeded (see Sections 1.4, 19.8, 19.9, and 19.10).

Single-pass fixed-tubesheet construction is common in shell-and-tube units designed to a maximum of 150 psi or that have a temperature drop of less than 50°F. Any larger drop limits the use of this construction due to thermal expansion considerations. This construction rarely has an odd number of tube passes unless that number is 1. Compare this with spiral-fin air cooler floating headers where connecting piping can accommodate thermal expansions and contractions, making it easier to have an odd or even number of tube passes. This variation applies to plate-fin exchangers with variations depending on built-in expansion capabilities. Knowledge of exchanger construction is necessary to make maximum use of the options available, as these affect geometry and cost.

This section applies to the tube side of shell-and-tube units, air coolers, plate-fin exchangers, double-pipe units, and others. The rates developed are based on the total outside surface of the tubes (bare or finned).

An exchanger selection begins with choosing the tube length and diameter. A length limitation, if there is one, is determined by plant standards, by shipping, handling, manufacturing, and cleaning considerations, by how well the exchanger fits in the available space, and by the surface required. If mechanical cleaning is needed, the design must account for it. The selection of tube diameter and gauge may be influenced by the data that follow. Two sets of examples are given, based on the fluid used, whether water or oil. For these the straight-travel pressure loss is assumed to be a constant so the effect that the tubing size has on exchanger geometry can be known.

There are other factors that affect tube-side pressure loss, including turning, entrance and exit, and expansion and contraction. These usually do not influence the heat transfer rate, though they do affect the head needed for selecting the pump. An exchanger's pressure loss can be known from installed pressure gauges as the difference in pressure at the inlet and outlet measured in pounds per square inch gauge (psig). Tubes are purchased by diameter (inches), thickness (inches), and length (feet). In the study that follows, exchangers have bare tube surface. Each is assumed to have a different tube diameter. The exchanger geometry will change on the basis of the tube chosen, as evidenced by the following.

Consider the tube side of an exchanger only. Each tube size is assumed to have the same pressure drop. All exchangers have 700 ft^2 of surface. In the first set of examples the assumed fluid is water; in the second, oil. The assumed water conditions are:

- Flow: 2600 GPM
- Straight-travel pressure drop: 2 psi
- Viscosity: 1.62 lbsm/ h · ft
- Thermal conductivity: 0.37 Btu/[(h · ft^2 · °F)/ft]
- Specific heat: 1.0 Btu/(lb_m · °F)

These conditions apply to all tube sizes. The second set of examples uses hydrocarbon-based oil to illustrate the same point. The assumed conditions for oil are:

- Flow: 3000 GPM
- Straight-travel pressure drop: 12.0 psi
- Viscosity: 36.3 lbsm/ h · ft
- Thermal conductivity: 0.074 Btu/[(h · ft^2 · °F)/ft]
- Specific heat: 0.48 Btu/lb_m°F
- Specific gravity: 0.9

The solution to the tube-side heat transfer rate curve, Figure 2-2, depends on the tube diameter selected as well as its thickness. The equation given in Figure 2-2 has been solved for many tube diameters and gauges using water or oil for the conditions specified. All solutions have the same surface (700 sq ft). The straight travel pressure loss used for water is 2 psi and for oil it is 12 psi. The results are given in Table 2-1 for water and Table 2-2 for oil and illustrate why one tube diameter and thickness may be preferred in terms of geometry and available space (columns 7 and 8). Changing a diameter can result in a change in the flow from laminar to the transition region to turbulent flow depending on conditions (Table 2-2). For simplicity the ratio of the bulk viscosity to the wall viscosity, μ/μ_w, is assumed to be one.

A good tube choice can lead to the most economical selection. In reviewing the tables, note the effect that the tube diameter has on these factors; some change more rapidly than others. The list includes tube length, tube passes, shell diameter, tube-side pressure drop per foot of travel, MTD correction factor, turbulent, transition, or laminar flow, and change in tube side velocity. Selective data has been taken from these two tables and restated for tubes that are all single pass and 10 feet long (Table 2-3) to emphasize the following points. The surface in shells of the same diameter and length is rapidly reduced as the tube diameter is increased. In applications where there is a high pressure loss per foot of travel the tube side loss can be minimized by using larger diameter tubes. While the tables refer to tube diameter, because that is the way tubes are ordered and exchanger layouts decided, the pressure loss that results depends on tube flow area and not diameter. On the shell side when the pressure loss is high, it can be reduced by increasing the flow area. No one speaks of increasing its equivalent diameter.

Some additional lessons are to be learned. In turbulent flow for the same pressure drop and surface, the tube-side rate varies little based on tube diameter. The tube length is shorter (for single pass) if small diameter tubes are used. For these the shell diameter increases assuming the tubes are single pass and on the same pitch. This conclusion changes as the number of passes increases. At this point the reason for choosing one tube size over another should be considered. Keep in mind that the smaller the tube diameter, the greater the surface that can be placed in a given volume (see Table 2-3).

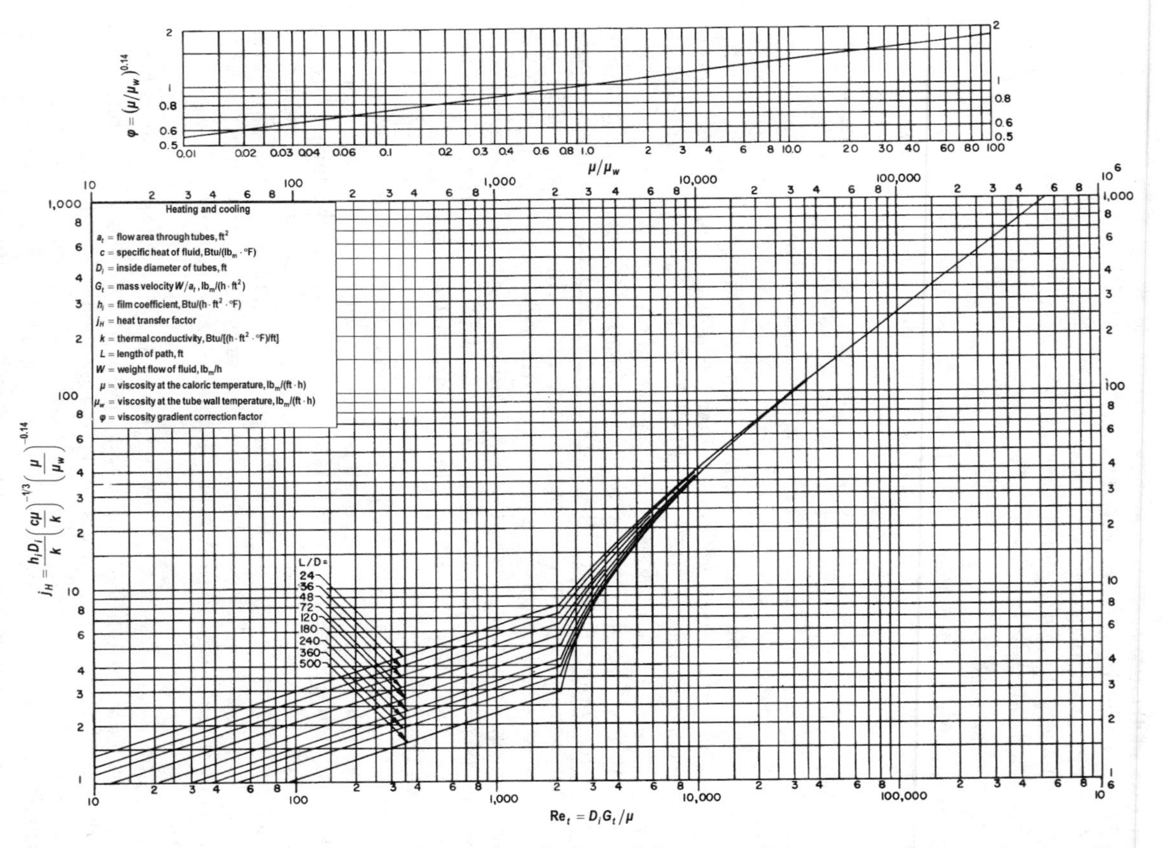

Figure 2-2. Tube-side heat transfer curve. (Adapted from Sieder, Tate, and Kern, *Process Heat Transfer*, McGraw-Hill, New York, 1950. Reprinted by permission of The McGraw-Hill Companies.)

Table 2-1 Variation in Exchanger Geometry Based on Tube Diameter and a Fixed Water-Flow Condition

- All selections have 700 ft^2 of external bare tube surface.
- Tubes spaced on triangular pitch, 1.25 times tube OD.
- Flow 2600 GPM and 2 psi pressure loss for all units.

Note: The volume of metal in the tubes does not vary much but the volume inside the tubes decreases as the tube diameter becomes smaller.

Tube dia, in	Gauge	Tube ID, in (nominal)	Film resistance based on tube OD, $\frac{h \cdot ft^2 \cdot °F}{Btu}$	Reynolds number Re	Water velocity, ft/s	Number of tubes	Tube length, ft	Shell ID for 1 pass tube side, in	Shell ID for 2-pass tubes: half the tube length; twice the no. of tubes, in	Vol. of metal in 18 BWG tubes only, ft^3	Vol. inside BWG 18 tubes only, ft^3
$5/8$*	14	0.458	0.00056	45,000	8.49	593	7.2	22	30.8		
	16	0.495	0.00055	47,100	8.23	526	8.1	21	29.0	2.628	6.46
	18	0.527	0.00055	48,980	8.06	474	9.0	21	27.0		
$3/4$	14	0.584	0.00056	57,000	8.45	368	9.7	21	28.0		
	16	0.620	0.00056	59,800	8.36	330	10.8	20	26.0	2.676	8.28
	18	0.652	0.00056	61,600	8.18	305	11.7	19	25.0		
1	14	0.834	0.00057	81,600	8.47	180	14.8	20	26.0		
	16	0.870	0.00057	83,800	8.34	168	15.9	19	25.3	2.770	11.85
	18	0.902	0.00057	85,400	8.20	159	16.8	18	24.5		
$1\frac{1}{2}$	14	1.334	0.00058	131,200	8.51	70	25.5	17.5	24.2		
	16	1.370	0.00058	133,400	8.43	67	26.6	17	23.6	2.769	19.14
	18	1.402	0.00058	136,000	8.43	64	27.9	16.5	23.0		
2	14	1.834	0.00061	180,500	8.52	37	36.0	16.8	23.7		
	16	1.870	0.00061	182,000	8.43	36	37.3	16.5	23.3	2.788	26.38
	18	1.902	0.00061	184,000	8.38	35	38.2	16.4	23.0		

*Pitch for $5/8$-in OD tubes is $13/16$ in triangular.

Table 2-2 Variation in Exchanger Geometry Based on Tube Diameter and a Fixed Oil-Flow Condition

- All selections have 700 ft^2 of external bare tube surface.
- Tubes spaced on triangular pitch, 1.25 times tube OD.

Tube dia, in	Gauge	Tube ID, in (nominal)	Film resistance based on tube OD, $\frac{h \cdot ft^2 \cdot °F}{Btu}$	Reynolds number Re	Water velocity, ft/s	Number of tubes	Tube length, ft	Shell ID for 1 pass tube side, in	Shell ID for 2-pass tubes: half the tube length; twice the no. of tubes, in	Vol. of metal in 18 BWG tubes only, ft^3	Vol. inside BWG 18 tubes only, ft^3
5/8**	14	0.458	0.011	2651	12.91	450	9.5	19.5	26.8		
	16	0.495	0.011	2761	12.52	400	10.67	18.4	25.3	2.628	6.46
	18	0.527	0.011	2884	12.35	360	11.85	17.3	24.0		
3/4	14	0.584	0.010	3369	12.91	278	12.84	18.3	25.9		
	16	0.620	0.0099	3443	12.44	256	13.95	17.4	24.6	2.676	8.28
	18	0.652	0.0097	3643	12.52	230	15.52	16.8	23.8		
1	14	0.834	0.0091	4827	12.95	136	19.64	17.3	24.5		
	16	0.870	0.0088	4995	12.85	126	21.20	16.8	23.8	2.770	11.85
	18	0.902	0.0085	5055	12.54	120	22.26	16.3	23.0		
1 1/2	14	1.334	0.0079	8203	13.74	50	35.62	14.7	20.8		
	16	1.370	0.0078	8324	13.59	48	37.10	14.4	20.4	2.769	19.19
	18	1.402	0.0077	8482	13.53	46	38.72	14.0	19.8		
2	14	1.834	0.0074	11,400	14.00	26	51.38	14.1	19.9		
	16	1.870	0.0074	11,300	13.47	26	51.28	14.1	19.9	2.788	26.38
	18	1.902	0.0074	11,500	13.53	25	53.44	13.8	19.5		

*Flow in 5/8- and 3/4-in tubes is in the transition range; in 1 1/2- and 2-in tubes, flow is turbulent; and in 1-in tubes flow is close to being turbulent.
**Pitch for 5/8-in OD tubes is 13/16 in triangular. Flow 3000 GPM at 12 psi pressure drop, all cases. Oil properties Section 2.4.

Table 2-3 Variation in Exchanger Geometry with All Tubes Shortened to 10 ft

Based on Tube Diameter and a Fixed Water-Flow or Oil-Flow Condition
(See Tables 2-1 and 2-2)

The volume inside the tubes does not change much with the tube diameter. However, the surface decreases rapidly with increasing tube diameter and so does the tube-side pressure drop. A high tube-side pressure drop can be reduced by using larger-diameter tubes.

Tube diameter		Water				Oil			
in	Gauge	Pressure drop in 10 ft of tubing, psi	Volume inside 10 ft of tubing, ft^3	Number of tubes	Surface, ft^2	Pressure drop in 10 ft of tubing, psi	Volume inside 10 ft of tubing, ft^3	Number of tubes	Surface, ft^2
5/8	18	2.222*	7.18	474	777*	10.12	5.45	360	590
3/4	18	1.71	7.08	305	598	7.74	5.33	230	451
1	18	1.19	7.05	159	416	5.39	5.32	120	314
1 1/2	18	0.87	6.86	64	251	3.10	4.93	46	180
2	18	0.52	6.90	35	183	2.25	4.93	25	131

*Pressure drop and surface exceed requirements because 10-ft tubes are longer than needed.

For the same tube OD, as the gauge becomes thinner, fewer tubes of longer length are needed for the same surface. Fewer tubes usually mean smaller shell diameters, less drilling, smaller flanges, thinner tubesheets, and other savings. For this reason the tube diameter selected often leads to the best economic choice. This technique is often used in selecting steam power plant exchangers where the use of costly nonferrous tubes (compared with steel) is common. Because of the quantities required and to reduce cost, orders for tubes are placed by specifying the exact diameter and thickness required including tolerances. The tube may be thinner than a standard tube gauge. In power plant applications the space available for feedwater heaters, including that for bundle removal, is usually defined. Historically, feedwater heaters are selected using 5/8-in and 3/4-in tubes to determine which fits best based on cost and space limitations. The thickness required is based on internal and external design pressure.

The pressure loss per foot will be less using large-diameter tubes. For this reason they are often used when flows are high (Table 2-3).

A properly selected tube size can result in a long tube that meets specifications. Long tubes generally result in lower costs. Large-diameter tubes in shell-and-tube exchangers generally increase cost because less surface can be placed in a given shell diameter (Table 2-3). Long tubes may require more baffles or supports to prevent vibration, but the cost of these will generally be less than the cost of exchangers with shorter tubes and larger-diameter shells, as previously noted.

Excessive tube length may be a problem. For example:

- The finished exchanger may not fit in the available space.
- Shop space must be available to complete the process of making U-bends. As an example, an 80-ft tube needs slightly less than 40 ft of open space to complete the bending process.
- Many plate-fin exchanger tube lengths are limited to 20 ft.
- If parts of an exchanger must be annealed, furnace length may be the limiting factor.
- The curvature of a railroad track sometimes limits the length of exchanger that can be shipped because of the resulting overhang that occurs when going around a curve.
- An exchanger that is too long will leave little space for mechanically cleaning tubes.
- Available tube lengths in seamless or welded metals are not necessarily the same. Make sure the length chosen satisfies your need.
- In air coolers the tube length is often determined by fan coverage.
- Over-the-road shipping limits can be the determining factor in selecting an exchanger.
- Excessive length can require that flagmen be used during shipment, possibly making the selection uneconomical.

For most of the above conditions it is good practice is to shorten the tubes and use twice as many, thus creating a two-pass exchanger. This idea may be repeated for four, six, eight, or other even-number passes. The preferred tube size is often the one commonly used in a plant and stocked. The following conclusions apply to multiple-pass tube-side exchangers, addressed by class of product.

2.4.1 SHELL-AND-TUBE HEAT EXCHANGERS

For shell-and-tube heat exchangers, counterflow is not always attainable, and an MTD correction may apply. The shell diameter increases with increasing tube diameter for the same surface and tube length. It decreases with increasing tube length. Thicker parts are needed as shell diameters operating at the same pressure increase.

As the number of tube passes increases, the complexity of pass partition plates increases. Small-diameter tubes require fewer passes due to the length lesson shown in Tables 2-1 and 2-2. Exchanger costs increase as more passes are needed.

2.4.2 Air Coolers

Header designs for air coolers include cover-plate headers (Figures 2-3*a* and 2-3*b*), shoulder-plug headers (Figure 2-3*c*), and U-tube or manifold headers (Figure 2-4). Some of these designs may not be the best choice for an application. For example, manifold headers cannot be mechanically cleaned, but they can sustain a higher pressure than other designs.

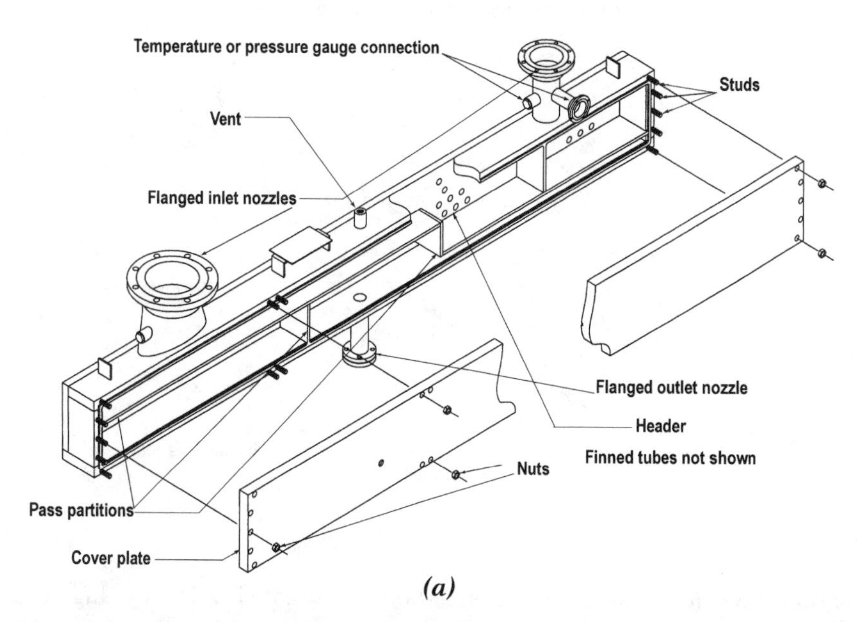

Figure 2-3. (*a*) Stud cover-plate header. (Reprinted by permission of Hudson Products Corporation.) *(figure continues)*

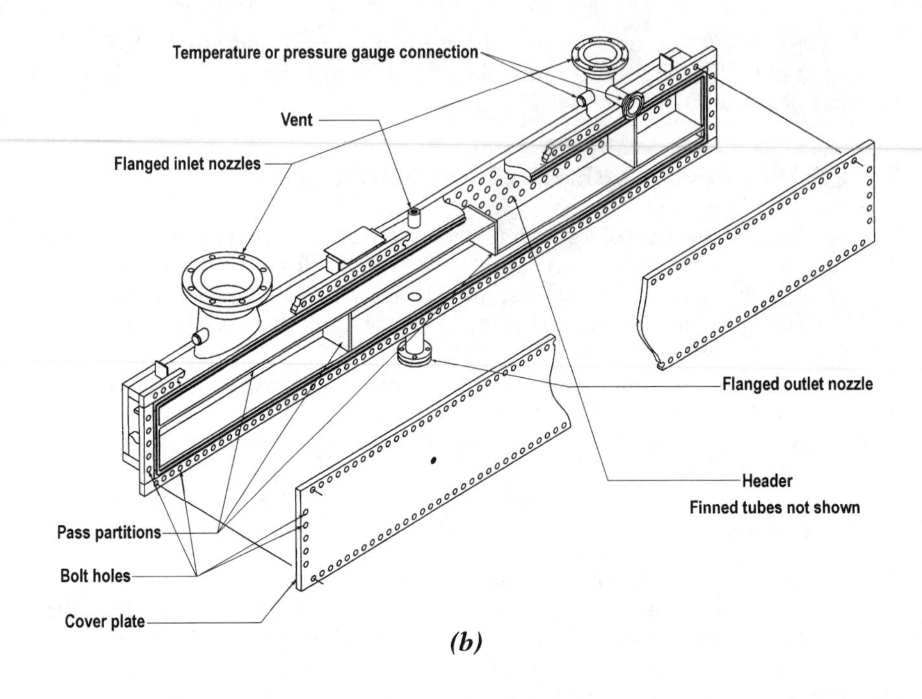

(b)

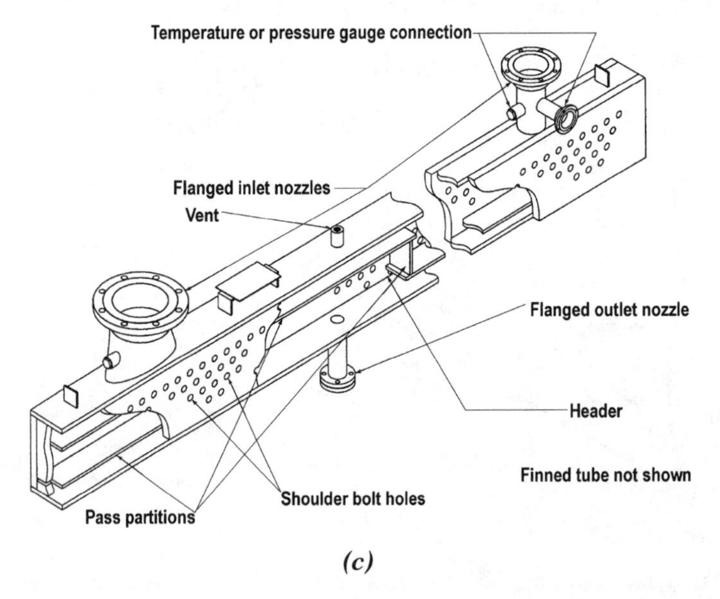

(c)

Figure 2-3. *(continued)* (*b*) Through-bolt cover-plate header. (*c*) Plug header. (Reprinted by permission of Hudson Products Corporation.)

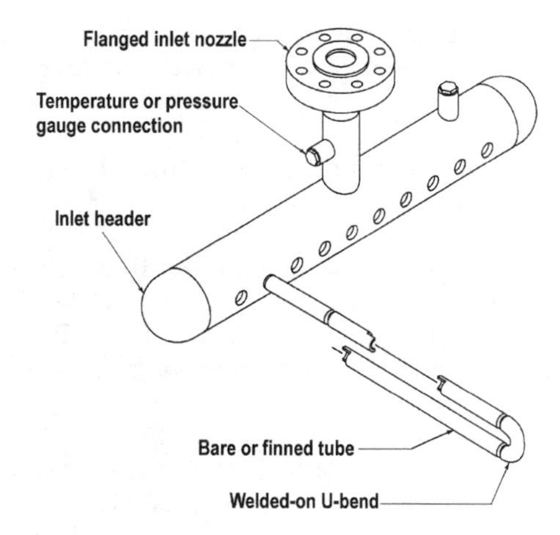

Figure 2-4. Manifold or pipe and U-bend header. (Reprinted by permission of Hudson Products Corporation.)

Cover-Plate Headers

The upper limit of the cover-plate header is usually about 300 psi and this header is in general use at operating pressures below this value. With a cover-plate header tubes can be mechanically cleaned. Horizontal or vertical partitions are easily installed. A practical limit is the complexity of gaskets and the ease of installing them, particularly when there are vertical partitions. More pass options are usually available using the cover-plate rather than the shoulder-plug design but these are limited by the operating pressure. If the needed value is above that given in Table 4.2 try a different selection or have the engineering department confirm that a design is acceptable. Keep in mind that Table 4.2 applies only to 1-in tubes on 2 1/2-in triangular pitch; different values apply to other pitches.

Box or Shoulder-Plug Headers

Box or shoulder-plug headers are good for negative pressure and for positive pressures of 1000 psi or more with minor modifications. The tubes can be mechanically cleaned. Horizontal partitions are preferred because they are easier to manufacture. Vertical partitions, in general, are to be avoided because of manufacturing difficulties. This sometimes limits the number of tube-side pass choices available.

Box and Cover-Plate Headers Compared

A substantial difference exists in the pressures that these header designs can sustain. Here's why. Consider a box header air cooler that has six layers and six passes, each row being one pass. After the fluid flows through one layer of tubes it passes into the opposite header where it reverses direction. The pass partition plate is welded to the header or box on four sides creating the needed seals. Welded in this fashion the pass partition plate also acts as a stay keeping the tube-sheet and plug header from bowing. In this case the unsupported length between layers is two.

The cover-plate header pass partition plate is sealed on one of its four sides by a gasket. It has no pass partition plate that acts as a stay to minimize header bowing. For the cover plate header in the above example the unsupported length between layers is six. Now refer to Table 4.2 in Chapter 4 and note that for a 3/4-in thick header the box header is good for 475 psi while the cover plate's limit is 70 psi.

Manifold Headers

This section discusses features of the manifold header heat exchanger design (Figure 2-4). Begin by reviewing Table 4.2 in the previous article and note that the design pressure that a header can sustain increases with a decreasing number of tube rows. The headers in this table are rectangular shaped. Manifold headers are round so they can sustain higher pressures, their principal benefit. The tube side can be designed for 15,000 psi or more. The inside of tubes cannot be cleaned mechanically. Even or odd numbers of passes can be furnished. This design almost always uses standard U-bends in terms of material, radii, and thickness to reduce cost and speed delivery. Tube spacing is wider, and so the heat of welding does not affect nearby fins (this subject is discussed in Section 4.2)

Manifold Headers of Brazed Construction

The manifold header design with U-tubes is the common construction of exchangers in the air-conditioning industry. Most metals used are non-ferrous, mainly of a copper base or of aluminum. This design can sustain high pressure but the corollary is also true. At lower pressures the relatively costly non-ferrous materials can be thin which is the usual case.

Seals between tubes and headers or between tubes and U-bends are usually made by brazing, hence the name, brazed construction. Brazed construction is usually limited to pressures of 300 to 500 psi, depending on construction details. With this construction tubes cannot be cleaned mechanically which is not a drawback for refrigerants because the fluids flowing are and should be in a sealed system. U-tube construction is almost always used.

The tube side of A/C evaporators and condensers should be capable of being drained so that lubricating oil can return to the compressor by gravity. The number of available passes is extensive because tubes are connected to each other using brazed connections. There is a large market for A/C products. For this reason a wide array of U-bends are available. For example, in a 60-tube exchanger, many combinations of passes and tubes per pass are possible:

1. 2 and 30
2. 3 and 20
3. 4 and 15
4. 5 and 12
5. 6 and 10
6. 10 and 6
7. 12 and 5
8. 10 and 6
9. 12 and 5
10. 15 and 4
11. 20 and 3
12. 30 and 2

In addition, the tube count could be modified to a 56-tube exchanger (8 and 7 or 7 and 8), a 63-tube exchanger (9 and 7 or 7 and 9), or a 64-tube exchanger with its variety of choices. Similar tube pass possibilities exist on the tube side of double-pipe exchangers.

2.4.3 Additional Lessons Learned from the Tables

The volume of liquid and the volume of metal in the tubes has been calculated for the different tube sizes used in Tables 2-1, 2-2, and 2-3. Some conclusions can be drawn:

1. The smaller the tube diameter, the smaller the volume inside the tubes for the same surface. In closed systems such as chillers, the reduced volume means that a lower quantity of refrigerant will be needed to operate the system. This can result in a large savings in refrigerant cost (this issue will be discussed in Section 15.1). The principle applies in other applications. An added benefit is that it can reduce the size of chiller or generator housings that have built-in exchangers.
2. In turbulent flow for the same thickness and pressure drop, almost the same quantity of tube metal is required for all tube sizes calculated. However, smaller-diameter tubes can withstand considerably higher pressures both internally and externally. This feature has the potential for considerable savings in tube cost, particularly if operating pressures are high. A good example is tubes in feedwater heaters. When selecting them it is usual to size these units twice, one unit with 5/8-in and the other with 3/4-in tubes. From these, select the lowest-cost unit.

By increasing the tube diameter, the Reynolds number can be raised from laminar or transition flow to the turbulent region with a resulting increase in rate (see Tables 2-1, 2-2, and 2-3). This has been known to raise the tube velocity sufficiently to be a negative in applications that use copper tubes. Copper pits when the water velocity is above 7 ft/s. This condition commonly occurs when the tube ID is increased. It rarely has an effect on tubes made of other materials.

The data in Tables 2-1 and 2-2 applying to 18 BWG tubes are rearranged in Table 2-3. All tube lengths have been shortened to 10 ft. Note that the volume inside the tubes, regardless of diameter, is nearly a constant. The pressure loss is less as the tube ID increases. More important, the smaller the tube diameter and spacing, the greater the surface that can be placed in the same shell diameter assuming the same tube length. As the tube ID increases, the surface inside the shell drops. Two lessons to be learned are:

1. For the same surface increasing the number of small diameter tubes in parallel decreases the pressure loss and the tube length.
2. Increasing the tube ID decreases the pressure drop and greatly reduces the surface in a given volume.

Costly errors in calculating pressure drop have occurred because tube and pipe tolerances were not considered. Table 2-4 lists several 1-in tube and pipe choices. Note the difference in IDs between tubes and pipes associated with their method of manufacture. The following gives an idea of typical manufacturing tolerances:

1-in seamless steel tubes	12.5%
1-in welded steel tubes	10.0%
1-in copper tubes	3.0%
1-in aluminum tubes	3.0%
1-in welded stainless steel tubes	7.5%
1-in seamless stainless steel tubes	10.0%

Tolerances vary depending on whether the tubes are minimum or average wall, of domestic or foreign manufacture, or hot- or cold-rolled. In refrigeration tubing, 1-in tubes mean 1 in nominal measured on the inside diameter. In process tubing, 1-in tubes are measured on the outside diameter. In the manufacture of plate-fin extended-surface exchangers, a further change occurs. Here the tube inside diameter is expanded about 3% to assure good contact between tube and fin. In the manufacture of lowfin tubing the tube ID and inside area are decreased substantially depending on the thickness of tube before finning began.

As the tube ID increases, the Reynolds number increases, and the rate can improve substantially if flow changes from transition to turbulent (see Table 2-2).

2.4.4 Plate Exchangers: Inferred Conclusion

Consider a plate exchanger having 24-in-wide plates 48 in long and spaced 4 mm apart. The equivalent diameter for this geometry is:

$$\frac{\text{Flow area}}{\text{Wetted Perimeter}} = 0.315 \text{ in}$$

The length of travel in a two-pass exchanger will be approximately 8 ft (refer to Table 2-1). A tube of this equivalent diameter is smaller than any selected

Table 2-4 Dimensions for 1-in Tubes and Pipes for Pressure Drop Calculations

Size	OD, in	Thickness	Mfg. tolerance on tube or pipe thickness*	ID, in	Notes
1-in steel pipe, welded or seamless	1.315	SCH. 40	10%	1.049 avg 1.0224 min 1.0732 max	1-in tubes and 1-in pipes have different outside diameters. Welded pipes may have a 7.5% thickness tolerance. Seamless may be as high as 12.5% thickness tolerance.
1-in type L copper refrig. tubing		14 BWG	3%	1.0 nominal	Refrigeration tubing is measured on the tube ID; the OD varies. Process tubing is measured on the tube OD; the ID varies.
Aluminum plate fins, copper tubing	1	18 BWG	3% on tube, 3% additional on expansion	0.927	Contact between the tube and fin is achieved by expanding the tube.
Theoretical tube	1	12 BWG	None	0.782	Theoretical dimensions are those usually used in calculating pressure drop and do not account for tube thickness tolerances.
Welded steel tubes	1	12 BWG min. wall	10% on welded	0.782 max 0.771 avg 0.760 min	Minimum-wall tube means the minimum thickness of a 1-in 12 BWG tube is 0.109 in.
	1	12 BWG** avg wall		0.802 max 0.782 avg 0.762 min	Average-wall tube means the average thickness of a 1-in 12 BWG tube is 0.109 in.
	1	11 BWG** avg wall		0.784 max 0.772 avg 0.760 min	For 12 BWG minimum-wall tubes the theoretical dimensions of an 11 BWG tube should be used. Note how close these dimensions are to each other.

*Varies by manufacturer.
**These tubes are usually considered interchangeable.

in the tube diameter study. The length of flow (8 ft) is on the short side. Hence, plate exchangers almost always have a higher allowable pressure drop per foot of travel and a shorter overall length of travel as well. Both conditions improve heat transfer rate(s) and make better use of the available pressure drop. For this reason higher rates can be obtained in plate exchangers than in most shell-and-tube units. The exception is small-diameter shell-and-tube units with short tubes, which have an even higher allowable pressure drop per foot of travel. The benefit of higher rates using plate exchangers (compared with large shell-and-tube units) has been confirmed by experience.

Chapter 3

Considerations in Selecting Shell-and-Tube Exchangers

3.1 GENERAL

Shell-and-tube exchangers are classified according to their geometry and construction. They have four main components: a bundle, a shell (with nozzles) enclosing the bundle, a stationary channel with nozzles, and the return head or its equivalent, U-tubes. Tubes are arranged within a circular area and rolled or welded into tubesheets. Shell-side baffles are positioned at intervals along the tubes to control flow, to act as tube supports, and to minimize vibration. Their function is to maximize the rate and stay within the allowable pressure drop. Meeting these conditions is not always possible in a given shell diameter, and when they cannot be met the shell diameter or tube pitch or both can be increased. Divided flow or the use of two or more units in parallel should be considered.

Many constructions are available. These and their nomenclature are given in the Standards of the Tubular Exchanger Manufacturers Association (TEMA), Figure 3-1. Shell-and-tube exchangers are classified according to construction features—principally, the stationary head, flow pattern, and rear-end closure. The latter is divided into three types: fixed tubesheet, floating tubesheet, and U-tube.

3.2 TUBE SIZES AND PATTERNS

In most exchangers the highest-cost item is tubing. Tube diameter, length, thickness, spacing, material, and quantity largely determine the lowest-cost unit. In the chemical industry the most common tube sizes are 5/8 in, 3/4 in, and 1 in OD. Refineries tend to standardize on 1-in OD tubes. These have proven to be economical and result in more compact exchangers than if larger-diameter tubes were used. Tubes larger than 1 inch in diameter are used when a low tube-side pressure drop is desired. The 1-in OD tube is beneficial when the tube fluid is fouling because this tube can be cleaned by mechanical means. Small-diameter tubes are not as efficient because of the greater number to be cleaned for the same surface. However, this construction is popular for other reasons: an exchanger with long tubes costs less on a dollars per square foot basis than one with shorter tubes.

Tubes are almost always arranged on triangular, rotated triangular, square, or rotated square pitch. Triangular pitch increases the surface that

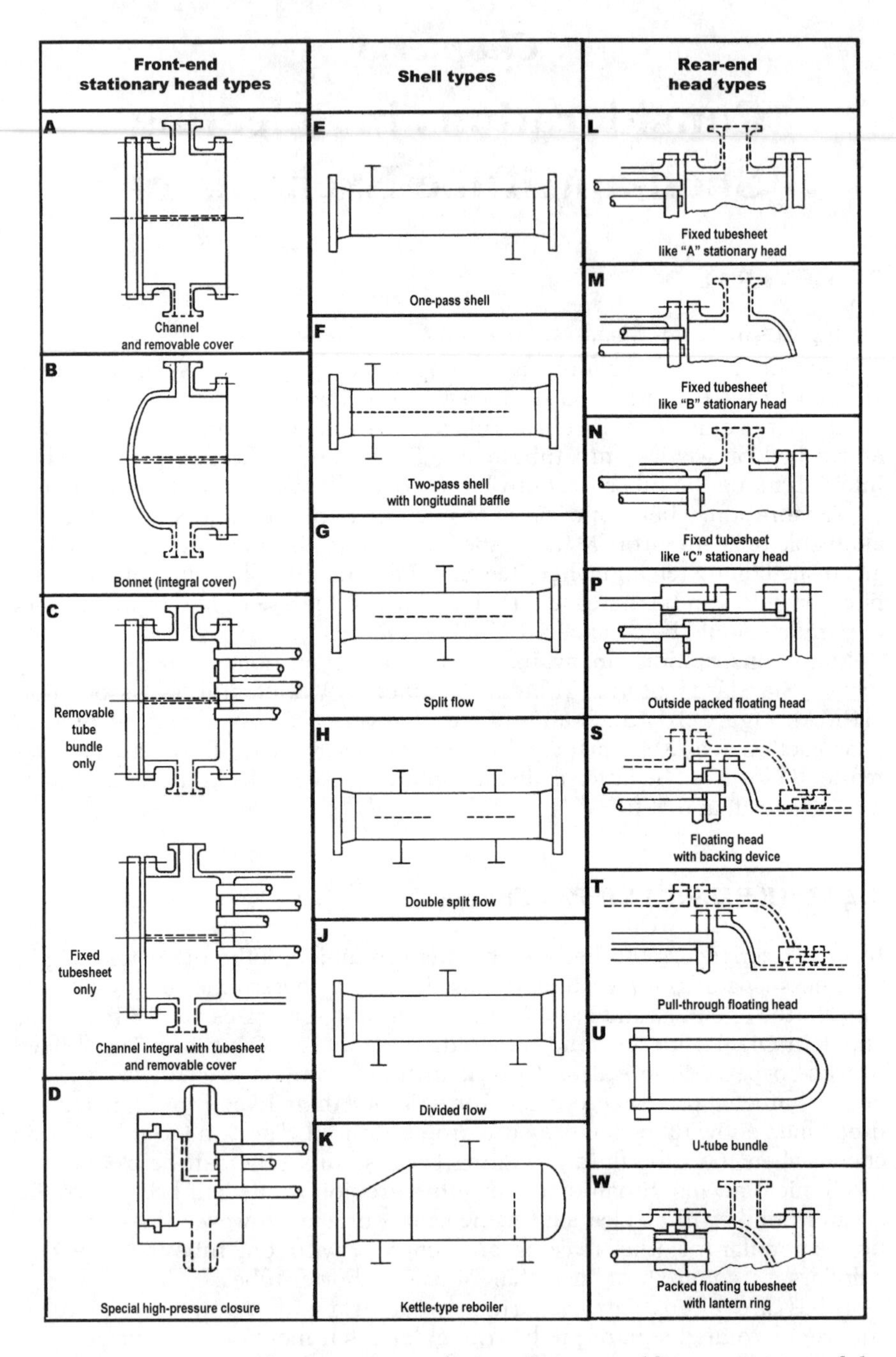

Figure 3-1. Heat exchanger nomenclature. (Reprinted by permission of the Tubular Exchanger Manufacturers Association.)

can be placed in a given shell diameter. Better shell-side rates are obtained using this construction. Square and rotated square pitch are preferred when the shell side is to be mechanically cleaned. This will be advantageous if a lower shell-side pressure drop is needed but will result in a lower shell-side heat transfer coefficient. Tubes are usually selected on the basis of the least expensive material acceptable for the job.

3.3 SHELL-SIDE DESIGNATION AND NOMENCLATURE

Several shell-and-tube designs are available. Their features are identified in this section to advise the reader of what constructions are available and to point out that the desired construction should be indicated in the specifications.

3.3.1 Fixed-Tubesheet Units (TEMA Type CEM)

The fixed-tubesheet exchanger consists of a number of straight tubes rolled into tubesheets enclosed by a circular shell. The clearance between the outermost tube and the shell is minimal; it is just enough for assembly. Tubes may completely fill the shell. Tubesheets are often made to a larger diameter than necessary so as to simplify construction at the channel flange. Typical TEMA designations for this construction would include BEL, BEM, CEM, BEN, and CEN (see Figure 3-1).

Fixed-tubesheet exchangers are often true counterflow exchangers. Counterflow reduces the surface required and the number of units, should a temperature crossover exist. The construction is simple, as there are no internal joints. Fewer tube-field bypass areas exist than for other constructions. The internal surface of tubes can be easily cleaned. This construction protects against leakage of shell-side fluids to the outside, as no shell-side gaskets are required.

The fixed-tubesheet unit is less expensive than other designs, but there are some disadvantages to its use. The shell side has limited access for inspection and can be cleaned only chemically or with steam, and the shell and tubesheets must be of materials that can be welded to each other. Muntz or naval brass tubesheets are not possible, because these metals cannot be welded to steel. However, these metals can be used as linings on the tube side of tubesheets. Tube and tubesheet materials need not be the same. In this frequently used design, provision for differential thermal expansion is minimal. A shell expansion joint is possible, but its integrity is influenced by the operating temperature and pressure, corrosion protection, start-up and shutdown conditions, upset or loss of flow in a stream, or presence of corrosive shell-side fluid.

3.3.2 U-Tube Exchangers (TEMA Type U Floating Head)

U-bend shell-and-tube exchangers are usually less expensive than other shell-and-tube designs, with the exception of fixed tubesheets. They have several advantages:

1. The number of rolled or welded tube to tubesheet connections is half that of a comparable straight-tube unit. Fewer connections reduce the potential for leakage.
2. A floating tubesheet and cover are not required. Eliminating these reduces the shell diameter and usually eliminates the need for shell and floating head cover flanges.
3. The bends usually accommodate thermal expansion and contraction forces, eliminating the need for a floating head.

Special costs result from the need to strengthen U-tubes at their bends, but these can be minimized. The bending of tubes into U-shapes during manufacture thins the tube at the centerline of the bend. This thinning becomes more severe as the bending radius becomes smaller and smaller. In the usual shell-and-tube exchanger application this thinning will be enough to require corrective action in the interior two or three rows of tubes of the bundle; these tubes will be too thin at the bends for containing the same pressure as the straight length.

A traditional way of correcting the weakness was to increase the thickness of tubes in the bend area midway in the tube length before the bending process begins. In effect the inner rows of tubes, before bending, are dual gauge with the thicker portion in the bend area. This method of manufacturing U-bends was once common, but few companies now make U-tubes this way. Instead tubes designated for internal rows are made one or two gauges thicker and, after bending, will be sufficiently thick to sustain the desired internal and external design pressure. The straight lengths of these inner rows of tubes only will be thicker than most tubes in the exchanger. There is no advantage to one of these tube bend designs over the other except that dual wall tubes often delay delivery time and some metals, with dual wall construction, are difficult to obtain.

Typical TEMA designations for U-tube exchangers include AEU, BEU, CEU, and AFU, among others (Figure 3-1). The U-tube design reduces cost because it eliminates the floating tubesheet, floating head cover, shell flange, shell cover, and bolting. In U-tube exchangers tubes are rolled or welded into the tubesheet. The clearance between the outermost tubes and the ID of the shell is minimal, being only enough for inserting the bundle into the shell. Tubes are spaced to fill the shell, although some available tube positions may be left vacant for specific reasons. There will be fewer straight tube lengths than if a fixed-tubesheet unit were used for the same shell diameter. For example, a minimum tube bending radius exists and results in space near the center of the bundle that cannot be filled with tubes.

Because only one tubesheet is required, it can be of any material that does not have to be welded to the shell. It need not be the same material as the shell or channel but often is if the metal is steel. Drilling time is less than for two-tubesheet units. Differential expansion is usually well distributed. Tube-side channels, gaskets, and covers and the inside surface of tubes are normally accessible for maintenance.

Very large temperature drops can be taken in one unit provided the hot stream is on the shell side and enters beyond the U-bends. Here the nozzle entrance loss is minimal and the dome area maximum. The temperature distribution along U-tube legs is nearly symmetrical using this construction. U-bends are advantageous in high-pressure applications, particularly when clean fluids are flowing.

There are disadvantages to using U-tubes. The mechanical cleaning of tube internal areas requires special tools. The outside surface of tubes not in the outermost rows cannot be mechanically cleaned when using triangular pitch. The number of tube-side passes must be even—not always advantageous. True counterflow occurs only when one fluid is isothermal. Replacing tubes, except for those in the outer rows, is difficult. Usually all tubes are replaced at the same time. The thickness of tubes at the U-bend of internal rows is the weak point of the exchanger, a condition corrected by using thicker tubes in these rows only.

3.3.3 Floating-Tubesheet Units (See Figure 3-1)

A floating-tubesheet exchanger consists of a number of straight tubes rolled or welded into tubesheets. One tubesheet is free to move, allowing thermal expansion between bundle and shell. Tubes do not fill the shell as completely as in the fixed-tubesheet or U-tube design. Space is needed for the floating head. Typical TEMA designations are:

- Split-ring floating head (AES)
- Pull-through floating head (AET)
- Outside-packed floating head (AEP)
- Packed floating head with lantern ring (AEW)

Split Ring Floating Head (TEMA Type S Floating Head)

Split-ring floating heads have several advantages. They can be used at higher pressures and temperatures than most designs. Tubesheets need not be of the same materials as the shell or channel, and tubes can be mechanically cleaned. Tube-side channels, gaskets, covers, and bundle are accessible for maintenance. Individual tubes are replaceable. Leakage between shell- and tube-side fluids is minimal.

There are also disadvantages. The split-ring floating head can be used only where a nonvisible internal gasket failure is tolerable. To remove the bundle, the shell cover, floating-head cover, and split ring must be removed. The distance between the shell ID and the outer tube limit is relatively large, and seal strips are usually needed (refer to Figure 19-8 in Chapter 19).

The number of tube-side passes is usually an even number. True counterflow cannot be attained except when one fluid is isothermal or the tube side is single-pass. Single-pass construction requires a packed head nozzle, thus increasing machining and assembly time.

Pull-Through Floating Head (TEMA Type AET)

The pull-through floating head has the advantages of the split-ring floating head except that the complete bundle can be removed without disassembling the floating head.

The disadvantages are the same as for the split-ring floating-head exchangers, with a few exceptions. The shell ID is larger and seal strips are wider, making this design more costly. There may not be a need for shell and shell cover flanges.

Outside-Packed Floating Head (TEMA Type P Floating Head)

Outside-packed floating-head exchangers have the advantages of split-ring floating-tubesheet exchangers except that they are lighter in weight and lower in cost, making them ideal for commercial service. Type P units have one principal disadvantage. Leaks at the packing gland will go to ground, creating a potential fire hazard if the shell-side fluid is flammable. For this reason these units are not acceptable in some applications in refineries and chemical plants.

The design is usually limited to shells under about 30 inches in diameter. Clearances between the outer tube limit and shell can be large, requiring seal strips in most cases. Shell-side pressures are limited to about 300 psi. This design is less expensive than most and is ideal when lightweight construction is acceptable and operating temperatures are below about 375°F.

3.4 SHELL TYPES

Many shell types are available, as shown in the middle column of Figure 3-1. The following subsections describe the advantages and disadvantages of the principal types.

3.4.1 One-Pass Shell (TEMA Type E)

The one-pass shell is a popular construction. Flow enters the shell at one nozzle and leaves at another. When only two nozzles are used on the shell side, this side is said to be single-pass. A variety of tube-side pass choices are available—that is, 1, 2, 4, 6, 8, 10, 12, or more passes—and all but single-pass are usually an even number.

3.4.2 Two-Pass Shell with Longitudinal Baffle (TEMA Type F Shell)

From a heat transfer point of view, the two-pass shell with longitudinal baffle is the same as two E shells in series. Heat transfer occurs across the longitudinal baffle, and this quantity is not always negligible, although it is most of the time. The advantage is that one shell rather than two will transfer the needed quantity of heat.

The disadvantage is in maintaining the seal between the longitudinal baffle and shell, particularly following removal and reinstallation of the bundle. Also, if there is too large a pressure loss between inlet and outlet nozzles, located opposite each other, it will place too heavy a load on the baffle, causing bypass to occur at the seals. Bundles have historically been difficult to replace while maintaining this seal. The flexible spring strip seal is more effective than most designs. The temperature drop in a shell should not exceed about 200°F. This design requires many quarter baffles, which add to cost.

3.4.3 Split Flow (TEMA Type G Shell)

Shell-side split-flow units have the advantage of one inlet and one outlet connection. Their principal advantage is that the shell-side flow splits into two streams with half going to the right and half to the left. Each stream has the same length of travel as in a once-through shell. Pressure drop is a function of mass velocity squared, and since half the flow goes in each direction, the straight-through component of pressure drop is one-fourth the loss that it would be through a single-pass shell. This construction has the disadvantages of the model E and F shells.

3.4.4 Double Split Flow (TEMA Type H Shell)

The double split-flow (type H) arrangement is a modification of the G arrangement (Figure 3-1) and is useful when long tubes are used. When the inlet nozzle is large, it can be replaced with two smaller nozzles, which do not require as large an inlet dome area. This means that fewer tubes are dropped for the dome area. The disadvantage of the design is that two inlets add to the cost of manifolding.

3.4.5 Divided Flow (TEMA Type J Shells)

The model J is used where minimal pressure loss is desired on the shell stream. This design is usually chosen for this feature. Typical applications include the following:

1. Services that have high vapor or gas flow on the shell side (either heating or cooling)
2. Gas or vapor services or a combination of these needing a low shell-side pressure drop (a heating or cooling service)
3. As a compressor intercooler with a moisture separator (a cooling service; some call this a TEMA type X shell)
4. As a compressor aftercooler with a moisture separator (a cooling service)
5. As a low-pressure nuclear power plant reheater with a moisture separator (a heating service)

Items 1 and 2 are bare-tube or lowfin shell-and-tube exchangers with segmental baffles and/or tube supports. The liquid or vapor or any combination of these enters the exchanger through one large nozzle or two smaller nozzles. In either case, half the flow travels through each half of the bundle. Fluid passage is in long and crossflow. Because the tube length is divided into two halves, the mass velocity through each half of the bundle is one-half that of the TEMA type E exchanger. The mass velocity is reduced to half in each direction, and the travel length is half the tube length. Therefore, the pressure drop will be one-eighth that of that of a once-through E shell. One drawback is that two inlets usually require extra manifolding.

Here is an example of where this construction is advantageous. Suppose the tubes are of 1-in OD on $1^1/_4$-in triangular pitch and the baffle spacing is 30 in. The liquid flow through a type E shell was high and the pressure drop was calculated as 35 psi. This is too high and can be reduced to under 5 psi by using a model J with two flow paths through the exchanger.

The type J or X is used in gas or vapor service that requires a low shell-side pressure drop, and it is here that fins are beneficial. In this design a tube bundle, similar to an air cooler section, is installed in a shell in order to contain pressure. The nozzles in some designs are installed on the same side of the shell to simplify compressor piping. One manufacturer's model J is shown in Figure 3-2. There are several reasons why this construction is beneficial. In some designs, two nozzles, each smaller than one large inlet, are positioned on the shell centerline and serve as inlets. They require a smaller dome area and will share the same one.

The length of travel across a finned bundle in crossflow is minimal. The pressure drop is lower than in the type E shell. The fluid crosses at the maximum area of the bundle and is almost always a single-pass design. The engineer should confirm that an MTD correction factor will not negate use of this design. This is a rare occurrence in pure crossflow units. The nozzles of model J designs can be reversed by using a single inlet and two outlets, but this means a larger dome inlet area will be needed and possibly a larger shell.

To better understand the model J, a review of exchanger features is useful. Consider the advantages of air coolers. The external to internal surface ratio is large. In unpressurized shell-side applications, the surface is at atmospheric pressure and air passes over a few layers of tubes. The short travel length and large flow area results in a low pressure drop. Air, the poorer conductor, is placed on the finned side, while the better conducting fluid (usually at higher pressure) is in the tubes. In coolers where air is not contained, some recirculates and mixes with incoming air resulting in a higher entering temperature and a possible performance problem. Neglecting this recirculation remnant, all air passes through the bundle one time. In compressor systems the same air passes through both intercooler and aftercooler, and bypassing is not a problem.

Assume the fin side of an intercooler and aftercooler must contain pressure. One way of ensuring this is to install a rectangularly shaped finned tube bundle in a shell. The bundle height is usually small relative to its

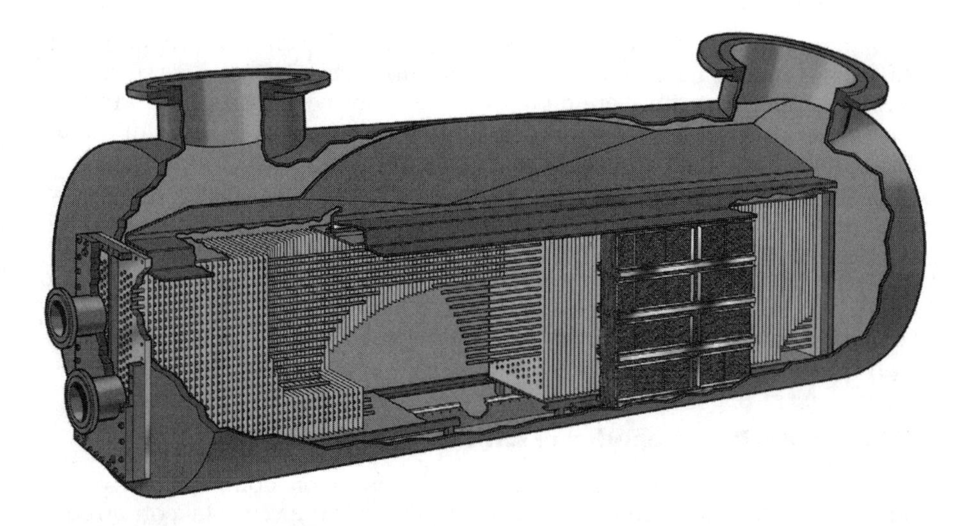

Figure 3-2. Cutaway view of model J intercooler. (Reprinted by permission of BDT Engineering, Industrial Products Division.)

width, though this is not a necessity. The shell diameter will be larger than the section width. The air-side rate will be improved as a result of increased pressure, while its surface and geometry offer the same benefits as at atmospheric pressure.

The advantages of this construction are.

1. Gases and vapors at low pressure have large-diameter pipes and nozzles to move the fluids and maintain a low pressure drop. The even larger-diameter shell simplifies the shell/nozzle connection and provides a large dome area.
2. The large dome area allows for good distribution across the bundle.
3. Condensation increases as air is cooled. The segment of shell at the bundle outlet is ideal space for a moisture separator. Drainage must be provided.

The model J is popular in intercooler and aftercooler applications. One reason is that it can deliver a low pressure drop in the first stage. If the second, third, and fourth stages increase the pressure 20 times, an additional 1 psi pressure drop in the first stage is equivalent to 20 psi less pressure delivered by the compressor. Compare this with the examples of critical pressure drop cited in Chapter 1 (screen room ductwork, Section 19.3; refinery output, Section 19.10; and chiller plant, Section 19.9). All underperformed because of too much pressure drop in the system. This is why a minimal loss in the first stage compressor cooler is necessary. The model J provides this and is the best reason for its use. It provides space for a moisture separator at a minimum loss of pressure, exactly what is needed.

A somewhat similar situation exists at the relatively low-pressure first stage takeoff in a nuclear power plant steam turbine. Here, the shell-side pressure and temperature are high, as is the vapor heat transfer rate. In this service the lowfin integral fin and tube with a high fins-per-inch count (up to 32) is preferred. This construction has minimal pressure loss in the first stage resulting in a higher-pressure output from the plant. It provides space for a moisture separator in the same way as the compressor intercooler and aftercooler designs. The specific heat of steam is about twice that of air. Half the weight of steam will be needed to transfer roughly the same amount of heat as air. Half the flow means the shell-side flow area can be less. This can be met by using a lower fin height and tighter tube spacing, plus the fact that high-pressure steam usually has a higher allowable pressure drop.

3.4.6 Kettle-Type Reboiler (TEMA Type K Shell)

The kettle reboiler is an E shell except that it has a large dome area for vapor separation. There are two exit nozzles. At the top is the vapor takeoff nozzle, and the nozzle beyond the weir is for liquid removal. Bundles are almost always removable.

3.4.7 Crossflow (formerly TEMA Type X)

The crossflow construction is beneficial in applications where the rating of the unit results in a short tube length, as in plate fin exchangers, or where the shell-side length of travel through a unit has to be short to attain a low shell-side pressure drop.

For good distribution, the inlet nozzle on steam condensers should be located at the midpoint of the bundle so that half the flow is in one direction and half in the other. This usually means a large dome area. The drain can be located opposite the inlet so that the shell can drain from two directions.

This construction is frequently used for air compressor intercooler and aftercooler applications. Shell-side air is under pressure, and the shell acts as the containing vessel. Tubes with high fins are beneficial. A large dome area is needed for distribution. The advantages are low shell-side pressure drop plus the benefits of finned surface.

3.5 FRONT-END TYPES

The types of front-end closures available are shown in the left-hand column of Figure 3-1. A description of the various types follows.

3.5.1 Channel and Removable Cover (TEMA Type A Front-End Closure)

Type A in Figure 3-1 is the popular type of front-end closure. The channel cover can be removed to inspect tubes without breaking piping. The tubesheets need not be of the same material as the channel. Removing the channel provides space for bundle removal and tube cleaning.

3.5.2 Bonnet (Integral, TEMA Type B Stationary Head)

The bonnet is much like the type A closure except there is no channel cover. Tubes cannot be inspected except by breaking piping and removing the channel head. The advantages are low cost, the reduced chance of leakage, and the elimination of a gasket.

3.5.3 Channel Integral with Tubesheet (Type C Stationary Head)

Being less expensive than the A or B closure, the type N construction can be used where channel and tubesheet metals can be welded. It eliminates one flange (channel to shell) and one gasket. The type C construction is similar to type N except that the tubesheet will be larger in diameter. This allows the metal in the increased tubesheet diameter to serve as a flange that will mate with a shell flange and gasket (not shown). The advantage of the type C construction is that the bundle is removable while in the type N it is not. In both designs the tubes can be inspected or cleaned by removing the channel cover.

3.5.4 High-Pressure Closure (TEMA Type D Stationary Head)

Exchangers that use the type D design, primarily high-pressure feedwater heaters, must be of the highest-quality construction, because power plants run continuously and their output affects many people's lives. The consequences of an outage are great. Plants often run continuously for five years or more without being shut down for servicing, so the quality of construction needed is the best available.

The type D head is a design for units that have high pressure in the tubes. Tubesheets and channels are often forgings. Pass partition plates (shown in dashed lines in Figure 3-1) are designed for the differential pressure across them, not the total pressure. High tube pressure tends to push the cover plate away from the pass partition plate, making it difficult to maintain the seal between them. Correcting this problem, should it occur, is difficult without this differential pressure seal.

Feedwater heaters require features not normally furnished with other shell-and-tube units:

1. Tubes and tubesheets can be subject to crevice corrosion. To prevent this in units where tubes are welded to tubesheets, the tubes are lightly rolled after welding almost the width of the tube sheet, that is, within 1/8 in of the shell side.
2. Feedwater heaters are designed for either vertical or horizontal installation. The choice is to make the best use of available space. Because high-pressure heads and channels are of thick materials, the weight of the bundle may be considerably above that of the shell. Therefore, for

Table 3-1 Properties and Uses of Shell-and-Tube Designs Based on Unit Diameter

Principal characteristics	Principal Applications	Limitations
Packed-head designs: 10 in or smaller in diameter		
• Tubes are usually in the range of 0.028 to 0.035 in thick. • Lowfin tubes are not often used here because the quantities needed are small. • The headers are usually cast iron. • Tube-to-tubesheet joints are seldom grooved. • Connections are usually pipe taps. • Many components are mass-produced. • Exchangers are mass-produced; few engineering changes are allowed. • In nearly all units a change in tubing material is permissible • These units have high heat transfer rates because of their high allowable pressure drop per foot of travel. • Typically, tube lengths are shorter than for process units. • Additional surface, when purchased new, is inexpensive. All that has to be done is to lengthen the tubes.	• Water-to-water exchanger • Water-to-oil exchanger	• Usually design pressure limited, shell 250 psi, tubes 150 psi • Usually temperature limit, approximately 350°F
Commercial designs: range to about 30 inches in diameter		

• Tubes are thinner than those used in ASME code units and usually thicker than those used in small packed-head units.
• Lowfin tubes are an acceptable alternative.
• Tube-to-tubesheet joints may be
  • Rolled only
  • Rolled with one tube-to-tubesheet groove
  • Rolled with two tube-to-tubesheet grooves
• Shell and channel may not require a corrosion allowance.
• The tube length is not standardized as in many refineries and chemical plants. Tubes are usually between 4 and 16 ft long.
• Baffle thicknesses can be thinner than for ASME code units.
	1. Light industrial applications
	2. Water-to-water coolers
	3. Lube oil coolers
	
	These units can be used in a great variety of industrial applications that do not require the demanding construction of ASME code and TEMA units. Standard drawings exist. Nozzles can be positioned to suit, and baffles can be spaced as required. The unit can usually be code stamped for a fee.
		1. Usually near these values (both streams)
		  • 300 psi
		  • 375°F
		  • 30 in. diameter
		2. Leaks to ground are easily identified. These exchangers cannot be used where fluid is flammable.

(continues)

Table 3-1 ***(Continued)***

Principal characteristics	Principal Applications	Limitations
ASME and TEMA units: no limit on diameter		
• Tube thicknesses are usually (not always) • 14 BWG min wall or thicker for steel tubes in chemical plants • 12 BWG min wall or thicker for steel tubes in refineries • 18 BWG min wall or thicker for nonferrous tubes • Tube lengths for refineries and chemical plants are usually standardized. • Shells and channels are carbon steel with 1/8 in or more corrosion allowance. • Tube-to-tubesheet joints are rolled and furnished with two grooves. • Welded (assumes materials can be welded). • Exchangers are designed for the application. There are no stock units. • The units usually require an ASME code stamp	High-pressure, high-temperature applications in refineries, chemical plants, and power plants or in applications that run continuously, i.e., 8760 hours per year. Their tube side can be lined with castables thus increasing the temperature at which they can operate. Safety is usually the defining term in their construction. Typically, these exchangers are not built in the same shops that assemble the packed-head or commercial exchanger designs.	There are limitations to ASME and TEMA shell-and-tube units. Some limits known to the author are: • Studies have been made of shells 50 feet in diameter; tubesheet and baffle drilling is the limiting factor. One unit built was 18 feet in diameter with 40 feet tubes. • A unit built had a tube side design pressure of 5000 psi. • A unit built had tube side design temperature of 1375°F (using castable linings this limit was considerably higher, near 1800°F). • Shell side design pressures were to 1200 psi. • The limits do not occur simultaneously. • When a large surface at high pressure or temperature is required, the usual answer is to furnish smaller units arranged in series or parallel or a combination of these.

ease of handling, shells rather than bundles may be removed to obtain access to the bundle. A head-up or head-down choice is based on ease of handling and overhead crane considerations.

3. The best designs have uniform steam distribution.
4. Tube vibration and drain control ("experience" related decisions) must be part of the design.
5. High-pressure closures must have tight joints. While the comment is obvious, delivering a unit to meet this requirement is not easy.
6. Zones: There are two kinds of feedwater heaters, those that operate at high pressure and those that operate at low pressure. High-pressure heaters require thicker tubes and tubesheets to handle the higher pressures. The pressure in the shells, while lower than in the tubes, is sufficiently high that the tubes must be checked for their design pressure in collapse. Feedwater heaters are often built with two separate and additional exchangers in one shell called zones. One, the desuperheating zone, makes use of available superheated steam to raise the temperature of the feedwater leaving the tubes to above that of the condensing steam. The other "built-in" exchanger is called the subcooling zone. It cools the condensate. Both designs make the system more thermally efficient particularly in minimizing the number of high pressure closures.

   Most shell-and-tube manufacturers do not make high-pressure feedwater heaters. However, the design of zones should be understood, because these can be useful in industrial and commercial shell-and-tube units.

3.6 SUMMARY OF SHELL-AND-TUBE DESIGNS AND THEIR LIMITATIONS

Three categories of shell-and-tube exchangers may be defined on the basis of unit diameter:

- Packed-head design (10-in diameter and smaller)
- Commercial units ranging up to about 30 inches in diameter
- ASME and TEMA units, with no limit on diameter

Table 3-1 lists the characteristics, applications, and limitations of the three categories of exchangers.

Chapter 4
Air Coolers

Before beginning the subject of air coolers it is a good idea to ask, Why should air be used for cooling rather than water? This chapter touches on a variety of subjects that might influence the answer. For a comparison of air and water as cooling fluids, see Table 4-1, "Air versus Water Cooling."

Air can be used to either heat or cool fluids and is available everywhere. There are more cooling than heating applications. Air is a poor conductor. To offset this it is beneficial to add surface to the outside of tubes that contain fluids to be heated or cooled. This improves heat transfer efficiency in terms of cost and size. Outside to inside surface ratios of 15 or 20 to 1 are common. Typical constructions include:

1. Spiral-fin surface (extruded, grooved, footed, edge-wound fins) added to the outside of tubes.
2. Punched holes in flat fins with tubes inserted in these holes. The tubes are expanded to make good contact between tube and fin. This construction is used in light services such as jacket water coolers, lube oil coolers, and A/C evaporators and condensers.
3. Lowfin tubes are beneficial when the shell-side rate is about one-third or less that of the tube-side fluid. Air is the usual shell-side fluid in air cooler applications. (Most lowfin tubes are in liquid service.)

The concern is to select the best surface for the application. The air stream can nearly always be single pass which will make maximum use of the available pressure drop for reasons that follow. In most applications the heat transfer rate of water will be roughly 100 times greater than that of atmospheric air. Due to this variation in rates, tubes that have high outside to inside surface ratios are ideal for cooling with air. Selections with this geometry will usually be competitive provided the available air side pressure drop is fully used. This can usually be met by changing flow area, the number of tube layers, the number of tubes per layer, the number of fins per inch, or the fin height and tube spacing. All these are controlled by the rating engineer.

There are major differences between air and water cooling. Air is often placed on the outside of bare tubes or those with fins; water is usually placed in the tubes. The water side will often require mechanical cleaning; the air side seldom does. If the tube side arrangement is changed from one to two pass, the tube side velocity will be doubled, its velocity component will be squared and be four times greater, and its length of travel is doubled thus increasing the pressure drop by eight times. Under these conditions all available pressure drop will seldom be used. Most air coolers have an allowable pressure loss near 0.5 in; this can be fully used by rearranging surface using

Table 4-1 Air versus Water Cooling

Subject	Comment
Auxiliary equipment	Air coolers do not require the additional cooling equipment that water-cooled units require. This includes cooling tower(s), pumps, valves, and water treatment facilities.
Availability	Water may not be available; air is everywhere.
Boiling	Unwanted boiling can occur on either stream of a heat exchanger depending on whether the low-boiling-point fluid is in the shell or tubes. The exchanger need not be "on" to experience this condition. Assume the ambient is 100°F and both streams are warmed 20°F by the sun's radiation. This heat will vaporize low-boiling-point fluids such as propane, ammonia, many refrigerants, etc., and cause a large increase in pressure in these streams, sometimes as much as 5 psi/°F of product temperature rise (refer to a properties of refrigerant R22, for example). This condition also applies to storage tanks containing fluids with similar properties, and applies to both water- and air-cooled units. (Section 18.11)
Cleaning	In general, access to the inside of tubes is easy. Process fluids or water will always be on the tube side of air coolers. Generally, a steam lance can be used to clean air cooler surfaces. Cleaning the shell side of shell-and-tube units and other exchangers can be costly, time-consuming and require considerable effort.
Coolant temperature	Air temperatures vary over a larger range than water temperatures. Air might be as low as –65°F in cold climates and +131°F in deserts. Water, by comparison, is usually in the 32 to 100°F range. Freezing can occur in either cooler but it occurs more frequently in air coolers. Freezing can occur in many ways.(See Section 12.) **1.** It can occur inside exchanger tubes when the ambient is below freezing. **2.** The moisture in outside air can freeze on the outside of tubes when the tube-side fluid is below 32°F. **3.** The moisture in make-up outside air (below about 25°F, usually) will freeze on the outside of tubes when this air does not completely mix with recirculating warm building air before crossing a bundle. Only one corner of the bundle may freeze, but the effect is the same as if a complete bundle froze.
Cooling capacity	Free circulation over warm tubes in air coolers provides some cooling even if power is lost.
Design temperature	Ambient temperatures vary considerably based on location. Typical design temperatures in the United States are 95 to 100°F; in Europe, 76 to 85°F; and in Canada, 65 to 85°F. These values vary and are not all-inclusive. Specific area design temperatures can be found in the latest edition of *ASHRAE Fundamentals*. Colder design temperatures mean smaller air coolers.

(continues)

Table 4-1 ***(Continued)***

Subject	Comment
Duct and line sizes	Refer to Section 20.4 and Chapter 27.
Elevation	This becomes a major factor when air coolers operate above about 3300 ft elevation. Adjust fan or blower horsepower and pressure drop accordingly.
Environmental	*Noise:* Air noise is more common than water noise, but, at times, water noise can be significant. Air noise can be reduced using a lesser fan blade angle, reducing fan tip speed, increasing the size of a duct, or adding sound-deadening insulation. *Water:* The effect of adding heat to a river or lake must be considered before the design water temperature can be finalized.
Fouling	The use of air for cooling eliminates one fouling stream because air is usually clean. Fouling on fins can be removed using a steam or water lance. Process fluids are in the tubes where cleaning is easier. Cover plates or shoulder plugs are removable for this purpose. Fouling on the shell side of shell-and-tube exchangers is difficult to remove. A large fouling factor in air cooler tubes tends to result in large units, strongly influencing exchanger size and cost.
Freeze problems	Freezing occurs more often in air coolers because of large swings in temperature. Shell-and-tube and other exchangers experience freezing as well. Fixes can be costly. (See under "Coolant temperature.")
Location	*Plant:* Water-cooled exchangers can be used where a source of water is available, while air is everywhere. *Air cooler:* Air coolers are usually not located near buildings except for small units so as to minimize recirculation. Consider installing them on roofs. If a ground-mounted unit is located near a driveway or vehicle traffic, a barrier will be needed to protect it from traffic.
Maintenance	In air coolers the mechanical cleaning of tubes and the servicing of drives are the main items requiring maintenance. Cleaning the inside of tubes requires removing the cover plate and/or shoulder plugs to obtain access. An all U-tube unit can, in most cases, be cleaned mechanically. Cleaning the inside of shell-and-tube exchanger tubes requires the removal of a cover plate as a minimum. It may also require the removal of a channel, bundle, floating head cover, split ring, or floating-head cover flange. Space for bundle removal must be provided. A crane may be needed for maneuvering it. The tube side of double-pipe exchangers can be cleaned mechanically by removing the tube connectors to obtain access. To clean the shell side requires space for removing U-tubes. Forced-draft air cooler drives can be serviced without impacting the process stream.

(continues)

Table 4.1 ***(Continued)***

Subject	Comment
Materials	Air cooler materials are usually selected based on the tube-side fluid. Air is usually noncorrosive, meaning aluminum fins can be used. The materials used in shell-and-tube, double-pipe, plate, and other exchangers must account for the needs of both flowing streams.
Noise	Refer to Sections 4.2, 4.3, and 17.3.
Outlet temperature	In general, the process outlet temperature will be 6 to 9°F colder when cooling is done by water rather than by air at design conditions. At other times the temperature could be colder or warmer depending on what ambient and water temperatures are at any given time.
Power outage	When power to an air cooler fan(s) is lost there will be a natural draft effect that takes place, much like a chimney effect, which provides some cooling. Water-cooled units merely shut down.
Pressure containment	Both sides of shell-and-tube and other exchangers are designed to contain pressure. Pressure is usually contained on the tube side of air coolers only, though the air side must be sealed to prevent bypassing.
Recirculation	Recirculation is more common in air coolers than in other types of exchangers. Sometimes unwanted water recirculation occurs. A common problem, the equivalent of recirculation, is to have warm water or warm air leave a neighboring plant and enter your plant or system at an elevated temperature. Don't overlook this when finalizing a plant design. (See Sections 20.1 and 20.3.)
Shipping	Air coolers are often selected based on a limiting shipping width. The next larger size may be too wide to ship without adding handling and wide-load charges. Another choice may be best. Shipping width considerations are less apt to be limiting for other types of exchangers.
Shock	Air coolers experience rain and hail shock (louvers are beneficial), which rarely affects other exchangers.

the methods noted above. Further, one pass cooling with air is common in air coolers but rare in shell-and-tube units. Thus, the eightfold pressure increase cited for tube side flow, is avoided. Increasing the width of an air cooler bundle adds air flow area and reduces the LMTD and surface required; increasing the diameter of a shell-and-tube bundle increases cost as shell containment parts become larger and shell diameters increase.

A good selection requires an understanding of exchanger geometry, heat transfer rate, pressure drop, and cost. Other factors are type of surface, air-moving devices, shipping, handling, installation, freezing, metallurgy, noise, tolerances, the cleaning method, thermal gradients, and tube diameter, to name but a few. Most of these are covered elsewhere in this book. For a discussion of air cooler geometry see Section 14-8, "Designs That Use Fewer Finned Tubes per Layer and More Layers," and Section 17.5, "Conserving Energy."

4.1 BEGINNING THE PROCESS OF SELECTING AN AIR COOLER

The question is, What type of air cooler is best for the application? In process work, as in other industries, the selection is an iterative one. Both initial and final selections should provide answers to questions similar to the following:

1. Can the equipment be shipped preassembled? When more than one bay is required, will the manifolding be symmetrical, to avoid distribution problems?
2. Is galvanizing the side channels required? Hot-dip galvanizing can be done before a section is assembled, not after.
3. Can the equipment be delivered without excessive shipping charges? The recommendation, in most cases, will be the economical one.
4. Is the surface adequate for the temperatures expected?
5. If a future expansion is planned, have concerns been addressed to simplify changes in distribution, manifolding, floor space, footing loads, isolation valves, minimum downtime, and similar areas?
6. Have alternative operating conditions been considered?
7. The most significant questions that specification writers must address relate to cost and maintenance: what is the nature of the application? If the unit is a jacket water cooler, can the usually lower cost plate-fin construction be used? Must two grooves be supplied at tube-to-tubesheet joints? Can thinner tubes be used? Is quick delivery a factor? What about cleaning? Should welded headers, cover plates, or manifolds be offered? If the cover-plate design is selected, will it meet pressure requirements? Will the cover be too heavy to handle during maintenance?

These issues should be addressed before finalizing a selection. The user should keep in mind that features wanted but overlooked in the writing of specifications cannot normally be added after equipment is built. Even a simple requirement such as adding an instrument coupling to a nozzle creates problems with the validity of an ASME Code stamp if the change is made after the stamp is applied. The lesson is, Develop the requirements before selecting equipment.

4.2 SOME GOVERNING CONDITIONS

Governing conditions should be known before selecting an air cooler. Several examples follow. In most, the least expensive exchanger is the governing condition. Even small coolers have limitations. The potential for problems is greater for large coolers, increasing the need for making the best selection.

A note to specification writers: Do not assume the engineer is aware of a governing condition unless it is in the specifications. All requirements should be in the specifications. The engineer can be notified of a late requirement by telephone, e-mail, fax, or letter.

4.2.1 Shipping, Field Assembly, Weight

The examples in Chapter 21 discuss shipping, assembly, and weight limitations that affect a selection. Other choices can usually be made that minimize these constraints and result in lower cost.

4.2.2 A Detail That Affects the Selection

Some process air coolers require welded tube-to-tubesheet joints. During manufacture the heat of welding is high, enough to heat nearby aluminum fins. If they are tension-wrapped, the heat will loosen the bond between fins and tubes locally. When welded tube-to-tubesheet construction is required, coolers with 1-in-OD tubes (a process exchanger standard), 5/8-in-high fins, and a tube spacing of 2-1/2 in are a commonly used construction. Distance between tubes using this spacing minimizes the effect that the local heat of welding has on loosening fins.

There are reasons for selecting one tube size and spacing over another. The shoulder-plug header can sustain high pressure, provided that pass partition plates are welded in place. Built this way, the plates act as stays, reducing the unsupported length of header while increasing the allowable design pressure. Tube spacing is a factor in tube support design. Supports are usually needed at intervals of 52 or 60 tube diameters (i.e., generally 52 or 60 in). Side channels are designed to keep headers, tubes, tube supports, and keepers within the channels so that louvers or plenums can be mounted and/or supported on the channel frames without interference.

Structural needs are so interrelated that changing a tube OD or fin height results in a new header design, new tube supports, a new design for bending the header, and new section drawings, all of which add to cost. There is an exception. The 1-in tube with 5/8-in-high fins can be replaced with a 1-1/4-in tube with 1/2-in-high fins. The OD of the fins is the same, allowing the same tube supports to be used. This design is beneficial when the tube-side pressure drop must be reduced. This is not to say that no other air cooler design exists. Designs with larger-diameter tubes and larger headers are standard with some manufacturers.

Tubes in plate exchangers have some support from continuous fins equally spaced along the tube length. These fins and tubes must have other supports at the same intervals as spiral-fin tubes. These bundles are also supported at 52 or 60 inch intervals. Plate fins are rarely used in process or high-pressure applications, because tubes used in this construction are relatively thin.

4.2.3 Noise

Low noise requirements can usually be met by reducing the fan tip speed. This results in less air passing over the coil. The LMTD will drop when this step is taken. This may add a tube row to be crossed or add tubes to a row effectively increasing surface. These changes are effective in lowering noise. (See Sections 4.3 and 17.3.)

4.2.4 Operating Pressure

For operating pressure limitations, see Table 4-2, "Estimated Maximum Allowable Design Pressure for Shoulder Plug Headers or Cover Plates." If the operating pressure is substantially higher, up to 15,000 psi, manifold headers can be used. Be aware that design pressures often determine the construction to use. Manufacturers of plate-fin air coolers rarely publish design pressure and temperature limits for their products because they vary with tube spacing. However, this information is readily available from the manufacturer.

4.2.5 Tube-Side Passes

Process coolers have one stream operating under pressure while the air side is at atmospheric pressure. In theory, there can be any number of tube passes, odd or even. In practice, the number is design-limited as illustrated by the following examples:

Table 4-2 Estimated Maximum Allowable Design Pressure (psi) for Shoulder-Plug or Cover Plate Headers

1-in tubes, 2-1/2-in triangular pitch, SA-515 gr. 70 material, corr. allowance 1/8 in, $E = 0.60$.

Number of tube rows	Header thickness, in				
	3/4	7/8	1	1-1/8	1-1/4
2	475	650	840	1050	1275
3	245	340	455	580	710
4	146	210	275	360	450
5	95	140	185	240	305
6	70	100	135	175	215

Notes:

- This table is for estimating purposes to determine if a cover plate or box header meets the pressure requirement. If not, change header geometry. Example: A three-layer section, unsupported, will meet a 215 psi design requirement using a 3/4-in-thick header. If a six-layer section, unsupported, is selected, its thickness will be 1-1/4 in, which is more costly.
- Metal thickness must be checked for each application, as the design limit depends on exact dimensions and tube spacing.
- The pressure that a section can contain increases as the unsupported length between rows decreases. Pass partition plates act as a stiffener or support for this purpose. This feature is not available using cover-plate headers. In other words, the more pass partition supports there are, the less the unsupported length between rows and the higher the operating pressure can be. Note that numbers and spacings of pass partition plates may be different (they usually are) at the inlet and outlet headers.

1. In a three-layer, three-pass section, one layer of tubes is one pass. This applies to a cover-plate, shoulder-plug, or manifold design. Inlets and outlets are at opposite ends of the cooler.
2. Now assume the same example, except increase the number of layers to four. The exchanger could be built using a cover-plate design. However, it is not good practice to select units with odd-shaped passes (each pass would have 1-1/3 layers of tubes), because of the difficulty of manufacturing and installing and aligning gaskets. Another selection should be made, or at least considered, to simplify assembly and maintenance.
3. If a shoulder-plug design is used, it will be difficult to build because it is not easy to reach the inner areas of pass partition plates to make vertical seal welds (though it can be done)(see Figure 2-3*c*). Manifold designs introduce another problem. Steel U-bends are usually available in only two stock sizes. Other U-bend sizes are nonstandard and must be special-ordered. For these reasons designs using standard parts should be chosen because of availability and reduced cost.
4. There is an exception to this: A/C evaporators and condensers, and applications that use copper tubes. For these an array of U-bend choices are available. Nearly all connections are brazed. The demand for this material in lightweight construction is large. New selections are usually not necessary, as many tube pass choices exist (see the discussion of manifold headers of brazed construction in Section 2.4).

4.2.6 Other Considerations

In fouling situations where a unit must run continuously and be cleaned mechanically, consider an exchanger with six sections, five active and one spare. An idle section could be cleaned while the others remain on line. This general idea applies to many applications. Its main disadvantage is the amount of additional equipment required, in this case, six inlet and six outlet valves plus their flanges, in order to isolate the sections.

When designing exchangers for current and future needs, symmetrical manifolding and identical sections, is the preferred construction. The question to be answered is, Should future manifolding needs, with blind blank-off flanges installed, be provided during the initial construction stage?

When mechanical cleaning is needed, the tubes should be of sufficient thickness (varies with material) to protect against damage during cleaning.

4.3 STRUCTURAL CONSIDERATIONS

Some process air coolers remove heat using air as the coolant. The goal is to select the best and least expensive cooler for the job. There are many variables that affect a selection, but a few dominate:

1. *Coordination of components.* Process air coolers are usually evaluated in groups rather than individually. A common structure built by one manufacturer and containing several services has been found easiest to handle. Sections are mounted side by side on the structure. In theory, one installation can and usually does have combinations of shell-and-tube, double-pipe, or plate exchangers built by many suppliers, but this is not true of air coolers. The advantage of a common structure favors one supplier furnishing all air-cooled units.

As an example of the necessity of coordinated design, pipe rack-mounted air cooler structures are designed to have a minimum number of parts. They must accommodate the bending moments of overhanging and service walkways, and the weight of sections, walkways, wind fences, recirculation chambers, louvers, wind loadings, manifolding, and stub columns. They must be simple to assemble and ship and be capable of being galvanized without warpage. They must accommodate brace locations that support drives and other components without interference. In northern climates these structures, walkways, and extra features can cost more than the cooling sections. Thus structure design is an economic factor in the selection of air coolers.

For simplicity in pipe rack-mounted structures, each section will nearly always have the same tube length to match adjoining sections and to simplify their header walkways (to service headers) and service and wing walkways (to service drives). This is why an order for all items is usually placed with one manufacturer. Ground-mounted air cooler structures have the same problems except that service and wing walkways are seldom required but header walkways are.

If the common structure, be it piperack or ground mounted, is to be galvanized, intermittent welding is not permitted. Continuous welding is needed for galvanizing. In the galvanizing process the structure can warp. Galvanizing causes structure warpage. The thinner the metal, the more likely that warping will occur. Straightening a warped structure or a piece of one is an expensive operation.

The benefit of a common structure, often required in chemical plants and refineries, is ease of assembly. Field assembly is simplified when adjoining bays and plenums are of similar construction. This applies to header walkways, service and wing walkways, stairways, sections that share common support column(s), and bracing. Column pads, in the main, will be alike. Similarly anchor bolt spacing simplifies the pouring of concrete bases and all can be poured at the same time. The painting or galvanizing of structures is simplified when similar requirements apply and the process is continuous.

Two common walkway types exist, grating and checkered plate. The grating walkway

costs more than the checkered plate type and is preferred in gas plants and similar applications because flammable gases cannot be trapped beneath them.

An order for many condensers and coolers, on a continuous structure, is common. As a practical matter, condensers should be selected first as these usually occupy about 75% of the space on the structure. The structure and its components often cost as much as the cooling sections. Selecting the right surface, minimizing structure size, fitting the coolers in the available space, be they ground or piperack mounted, at lowest cost is the challenge facing the rating engineer.

2. *Space.* Air coolers occupy large areas. The preferred selection must be adaptable to the available space.

3. *Noise.* Noise is a major consideration. Fans, speed reducers, and motors have moving parts that generate noise. Structural vibrations caused by source currents increase noise levels. The degree to which fan speed, blade angle, clearance, and airflow are controlled affect an exchanger selection. In general, a slightly larger exchanger using a fan operating at a lower tip speed has a lower noise level than one selected based on heat transfer alone. Reducing fan speed is one of the better ways to reduce noise. (See Section 17.3.)

4. *Assembly.* Field erection is expensive, which is why maximum shop assembly is preferred. Avoiding the marking of parts reduces handling and costs. When large air coolers are shipped knocked down, manufacturers should consider the use of sea-land containers, because units boxed and shipped this way are less likely to arrive with parts lost. Lost parts during shipment are a trouble job of a different kind.

5. *Power consumption.* Power should be part of cost evaluation. A cooler with a wide structure can be the lowest-cost exchanger because the power required for operation is less. (See Section 14.8.)

6. *Wind.* Wind shifts and recirculation affect the performance of air coolers—a relatively uncommon occurrence for water cooled exchangers. (Recirculation is discussed further in Chapter 20.)

7. *Interference from surroundings.* The location of equipment relative to buildings, cooling towers, plant services, etc., can affect performance. (See Chapter 20.)

8. *Number of connections.* Manifolds can be lengthy, with many connections. Engineering these adds to design time. The use of many connections is expensive. While manifolds, technically speaking, are not part of the air cooler, minimizing the number of connections is part of a good selection. (Manifolds and headers are discussed further in Chapter 11.)

9. *Suitability for location.* Units that are recommended for service at one location may not perform at another, because of differences in the design temperature or elevation.

10. *Competing products.* Part of the decision of choosing an air cooler is evaluating the offerings of several manufacturers for the same service.

Review these variables and note that most are independent of heat transfer surface, which is the rating goal. The least expensive surface is almost always the one with long tubes. Adding length increases surface but does not increase tube hole drilling, alter header construction, or change piping, but it makes finning more efficient. Sometimes a manufacturer's location governs its ability to ship economically. The lower 48 states do not all have the same shipping width limitations. The need for flagmen in some states further affects the economic choice. Hence, there are times when a plant's location makes its product noncompetitive because of shipping charges.

At this point, conditions have been noted that affect the cooler's geometry. Because walkways are easier to install in a straight line and because pipe rack centers affect the design of the structure, the economics of an installation begins by selecting a common tube length for all sections. Now to the equipment.

Air-cooled exchangers differ from others. They occupy more space, cost more initially, use substantial amounts of nonferrous materials, have moving parts, present shipping problems, and often require field assembly. Their construction must be understood in order to make the best selection for each application.

The minimum air cooler consists of a section or coil, structure, fan, speed reducer, and driver. Other equipment might include louvers, a recirculation chamber, steam coils, ladders, walkways, autovariable fans, cutout switches, wind fences, and noise abatement features. Forced- or induced-draft construction is available. Consider two coolers with the same surface; the only difference is that in one the surface is arranged in six layers, and in the other, four. A four-layer section has a 50% wider structure. Reread the beginning of this paragraph, on air cooler components and auxiliaries, and picture how the equipment width will increase using a four-tube-layer arrangement by increasing the width of bay, size of louvers, wind fence, and section heaters, diameter of fan, and size of drives. As a plus, the four-layer arrangement reduces the energy cost of moving air because the air flow area is increased and the number of tube rows to be crossed is reduced from 6 to 4.

A heat exchanger manufacturer's responsibilities end at inlet and outlet flange faces. In air coolers they also end at the structural base plates. A user handling problem sometimes occurs when the user's engineers design footings without considering future plant expansion loads. Footings are relatively inexpensive. Correcting them for future loads is very expensive, primarily because one may have to shut down a plant to make the correction (see Section 21.9).

4.4 FORCED-VERSUS INDUCED-DRAFT AIR COOLERS

Air coolers can be built of either forced- or induced-draft construction (Figure 4-1). One is often preferred over the other. There are two types of each, those that stand alone and those that share structural columns with other air coolers. A stand alone coil or heat exchanger section with one fan mounted horizontally and with cooling air moving upward has four vertical sides or areas where air can enter to provide the desired cooling. If the cooler bay has two fans, air can approach each fan on three sides only. If the bay is mounted next to a wall or building, air can reach each fan on two sides only; mounted in the middle of a bank of exchangers, the flow area is from one side only. Hence, when selecting either type of air cooler it is important to know where a coil will be located relative to other coils, walls and buildings.

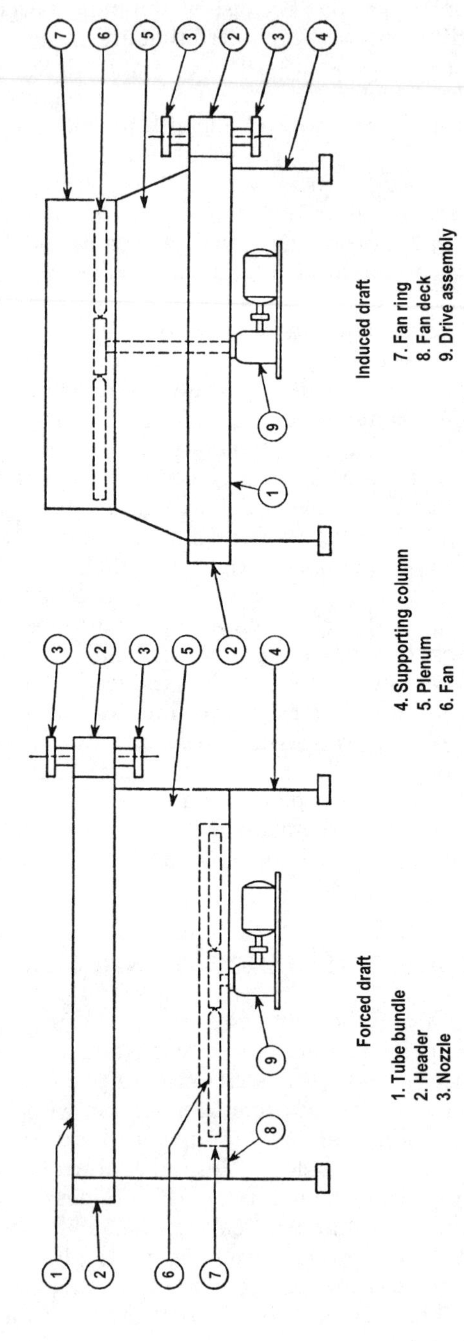

Figure 4-1. Typical components in an air-cooled heat exchanger. (Reprinted courtesy of the American Petroleum Institute.)

If not accounted for, starving fans is possible. To calculate opening sizes, see Chapter 20. There is an exception when pipe rack mounted units are used. Cooling air can be obtained from beneath the coil(s) provided they are mounted a significant distance above grade.

The advantages of forced-draft air coolers follow:

1. The air-moving equipment is mounted in the cold stream and so is more adaptable to more applications, particularly when air discharge temperatures exceed 160°F to 180°F. At these temperatures the life expectancy of mechanical equipment will be less. Special attention must be given to the materials of construction and to lubrication when drives operate in high-temperature surroundings.
2. Less horsepower is needed to move air through the coil using the forced-draft design because the air to the fan will be colder.
3. Forced-draft units can be made with air recirculation built in, a feature used to prevent freezing in cold weather.
4. Inspection of drives (Figure 4-2) is easier because they are more accessible. In general, it is not necessary to turn off the tube-side flow to maintain the fan, motor, reducer and V-belt drive(s).
5. Mechanical equipment is slightly less expensive and can be supported independent of the structure if necessary. Using the forced-draft design means less vibration is transmitted to the exchanger and structure. The equipment is positioned in such a way that it is easier to maintain.
6. Bundles can be removed without disturbing the structure or mechanical equipment.
7. Forced-draft designs are beneficial for economic reasons where two services can be combined in one cell. This is equally true of induced-draft units.

The advantages of induced-draft air coolers are:

1. The air leaves the cooler at a higher discharge velocity, lessening the tendency for recirculation to occur. The induced-draft design is best near buildings, where a tendency for air to recirculate exists.
2. The shroud offers partial protection from rain, hail, sun, and flying matter (sand, leaves, etc.).
3. Improved performance can be obtained by adding a recovery stack. When this is done, the fan horsepower can be less than it would be if the forced-draft design were used.
4. In the event of a power failure, partial cooling is available with the fans off, but this is equally true of forced-draft units.

4.5 PROCESS HEAT TRANSFER COIL OR SECTION

Most heat exchanger inlet and outlet nozzles are fixed in position. If thermal expansion or contraction occurs, it must be accounted for in the piping. In process air coolers, some movement can be designed into the coil to

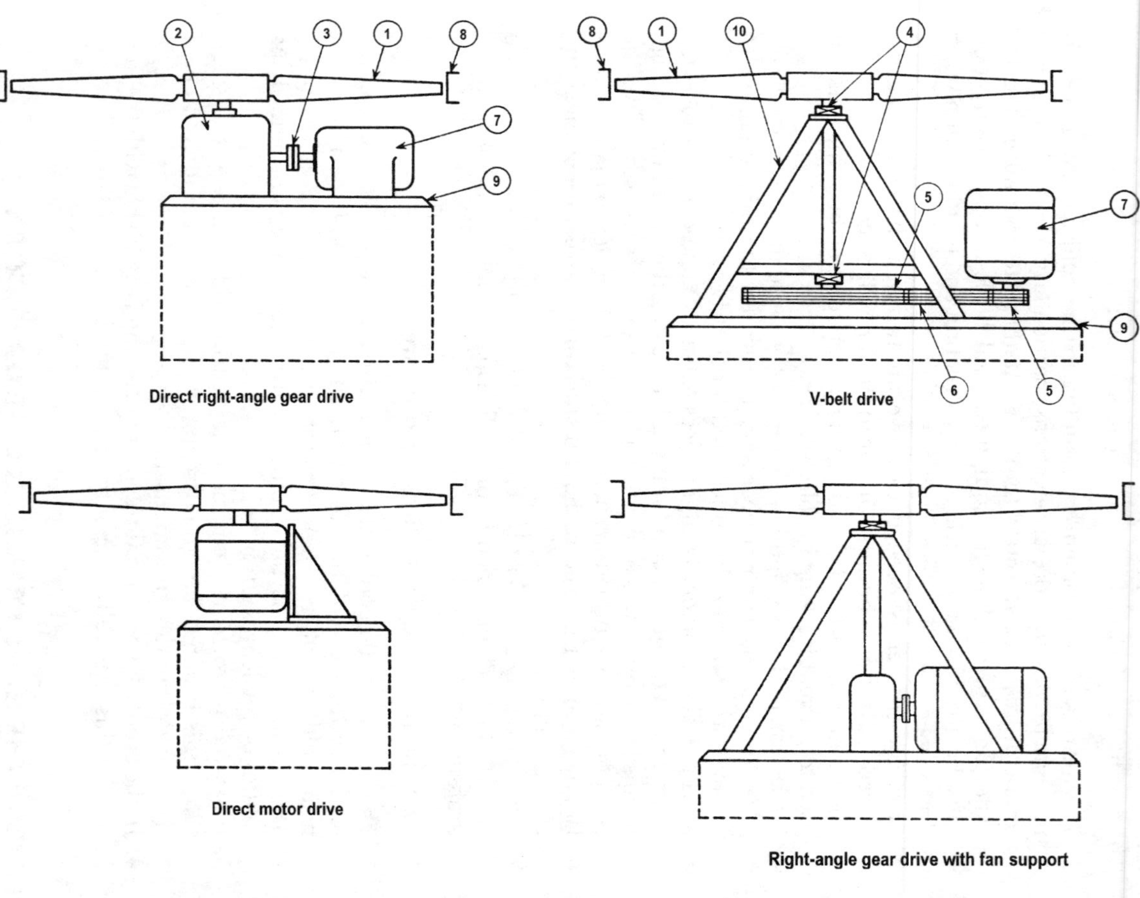

Figure 4-2. Typical drive arrangements. (Reprinted courtesy of the American Petroleum Institute.)

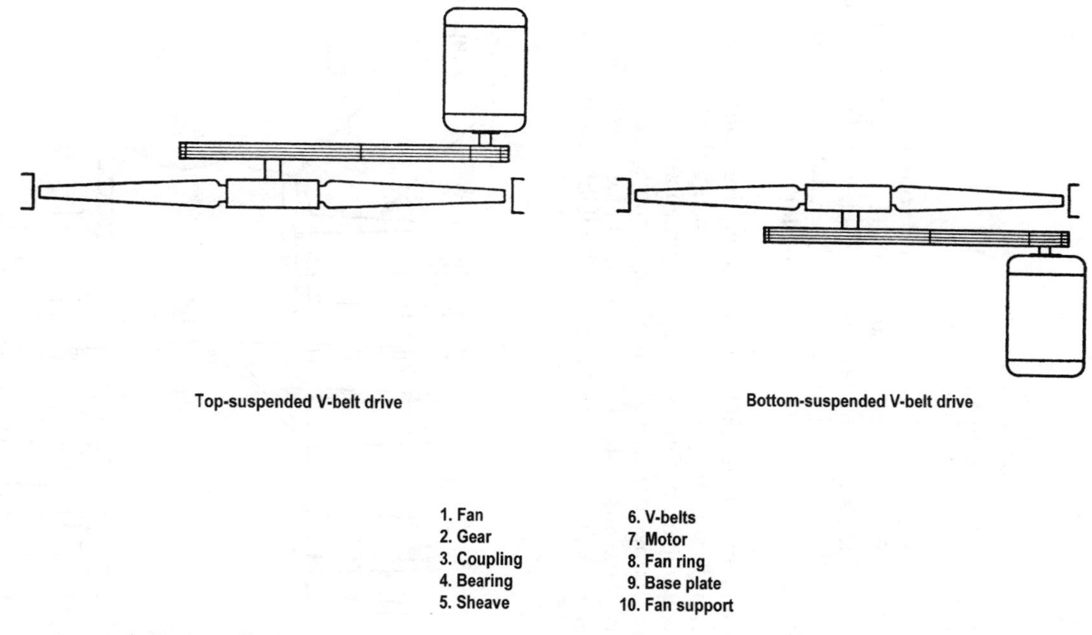

Figure 4-2. *(continued)*

accommodate expansion and contraction forces. Other features can be built into air cooler sections, and some of these follow. Keep in mind that if the features are a requirement, they must be included in the specifications.

Using aluminum as the fin material, the surface in process coolers can be constructed in four ways:

1. Spiral fin, tension- and edge-wound
2. Spiral footed fin, tension-wound (Figure 4-3)
3. Spiral fin, edge-wound into a groove (Figure 4-4)
4. Extruded aluminum fins (Figure 4-5)

Other fin materials are available but are not used as often as aluminum in air cooler applications (Section 5.1) . Tension-and-edge-wound spiral fins are also available in copper, steel, stainless steel, nickel, nickel alloys, and most non ferrous alloys. These can be used with nearly all tube materials. The reasons why they are not is poor fin thermal conductivity (copper and aluminum excepted), and thicker materials are needed to prevent the tearing of fins during manufacture. Bending these metals requires considerable energy. Thicker fins mean fewer fins per inch thus increasing the number of tubes required for an application. Fins of the spiral footed and extruded types are available in aluminum, copper, and some copper alloys.

Many metals are available for fins edge wound in grooves. This construction is useful as design temperatures increase. Available fin materials include

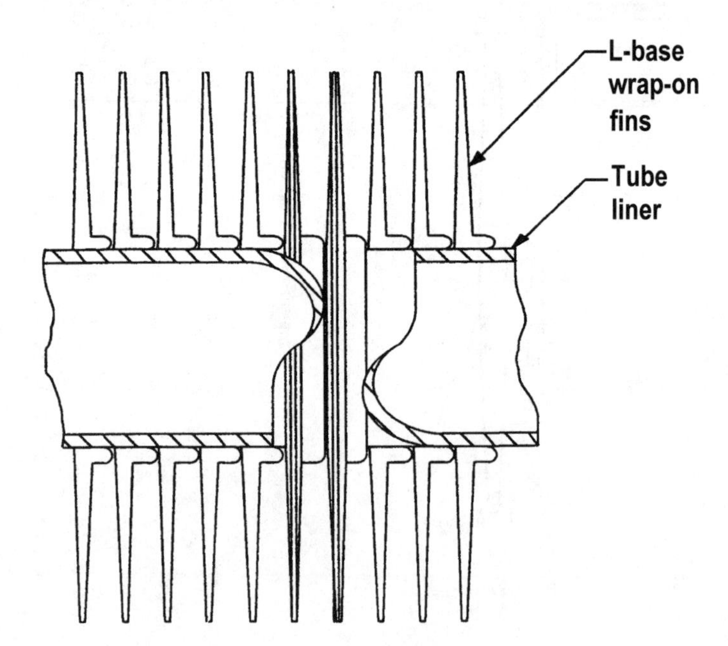

Figure 4-3. Spiral footed fin, tension-wound. (Reprinted by permission of Hudson Products Corporation.)

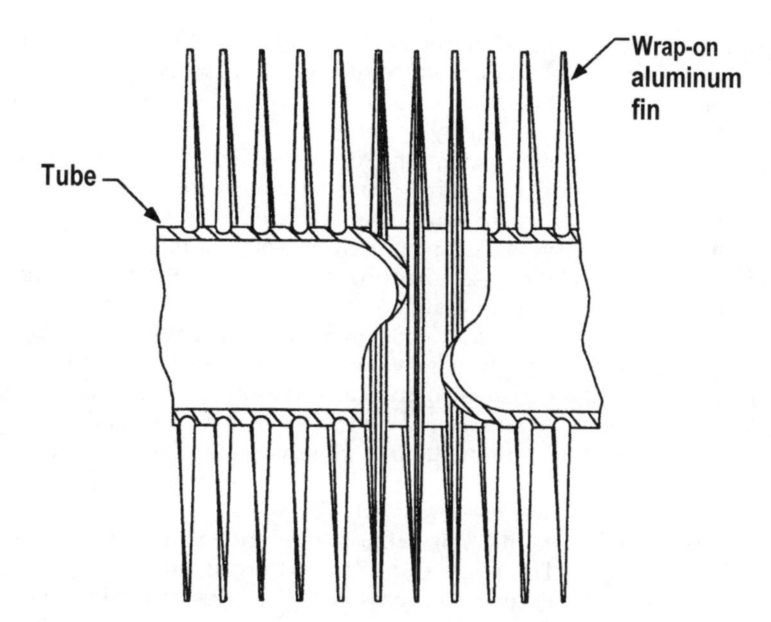

Figure 4-4. Spiral fins edge-wound into a groove. (Reprinted by permission of Hudson Products Corporation.)

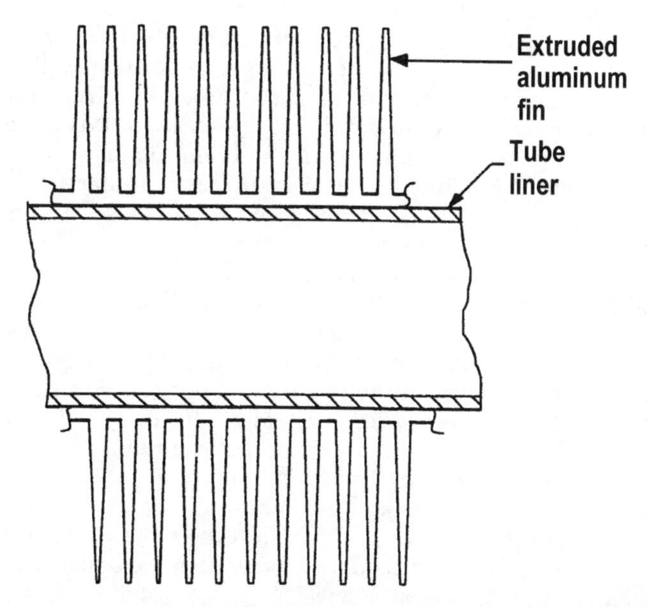

Figure 4-5. Extruded aluminum fins. (Reprinted by permission of Hudson Products Corporation.)

aluminum, copper, steel, stainless steel, nickel, nickel alloys, and possibly others. Not all manufacturers produce fins of every material. In most cases, a change in fin material results in a change in the tube's design temperature plus other limits such as fin height, number of fins per inch, fin thickness, and availability to name a few. If, for example, one wanted a hastelloy fin, a manufacturer might make a sample tube of this material to confirm that construction is feasible before committing to building it.

Edge-wound spiral fins are wrapped tightly around a tube and seated using knurling rolls. The process raises the tube metal against the base of the fin to form a bond. The bond has its limitations and can unravel at fin ends under unfavorable circumstances. At fin ends, provision must be made to keep this from happening. In general, the edge-wound fin is the least costly construction.

Fins with a footed base can also be wrapped around a tube, when it is preferred that the bare tube be covered by the fin feet. Pressure, designed into the finning process, forces fin metal into close contact with the tube, making the bond more secure. Provision must be made to keep fins designed this way from unraveling (Figure 4-3).

The grooved edge-wound spiral fin is constructed by plowing a deep groove into the tube. The edge-wound fin is forced into this groove. The metal displaced in the grooving process is forced against the base of the fin to form a metal-to-metal bond. Fin unraveling is rarely a problem. This tube can operate at higher temperatures than other designs (Figure 4-4).

An outer tube of thick aluminum (there need not be an inner tube in some cases) can be modified to form extruded fins. A strong mechanical bond between the fins and tube is formed during this process. The design is competitive with wrap-on and footed fins and can be used at higher operating temperatures (500 to 550°F), depending on the tube wall temperature. In some cases, such as condensing steam in tubes with little fouling, the upper operating temperature limit can be easily reached. Process and metal temperatures are nearly the same, with the bond temperature limit dependent on inner tube metallurgy. Check with the manufacturer regarding the extended fin's suitability for your application. This fin almost completely covers the tube except for small areas near tubesheets. This is a plus, particularly in corrosive atmospheres (Figure 4-5).

Well-designed process sections can have the following features. Provision for thermal expansion of the floating head is nearly always required. The floating head may experience a problem during transport, as longitudinal shock loads will be transmitted to the tube-to-tubesheet joints. Failures at tube-to-tubesheet joints are common when this occurs. To prevent this from happening, shipping plugs or welded-on clips are installed to hold the floating head in position during transport. The plugs and clips are designed to take transport shock loads and are removed when the sections are in position and before plant start-up. Other fin-tube types exist, such as those that are hot-dipped galvanized, soldered, brazed, or welded and those that are of other materials, to name but a few.

Allowance for lateral expansion or contraction at the inlet end of the exchanger can be part of the exchanger design, but it must be requested in the specifications. Movement may be up to 1 in, or any combination, plus or minus, that adds up to 1 in. The complete section does not move, only the inlet header. This feature may simplify piping.

Engineering should approve any selection that has more than a 50°F temperature drop per pass. All sections with this or a larger temperature drop in the last pass require split headers that are usually one layer deep. Choose coolers that have no wide-load or flagman charges, for economic reasons. Try avoiding these fees, as they are cumulative when travel is across several states. In effect, width determines the economic choice of coil, structure, and air cooler. Wide sections and additional shipping charges usually result in a noncompetitive selection. Sections that can be shipped without wide load and flagman can usually be preassembled, thus avoiding field assembly charges. If wide loads are within shipping limits, there is little a rater can do to reduce shipping charges. The economical width varies with geography.

Generally speaking, wider sections are more acceptable the closer the installation site is to the manufacturing facility because there are fewer road and bridge limits to contend with. In process work, the best selection usually has two V-belt drives per bay and large-diameter fans. Sections should be designed with no loads on the nozzles during shipment. Little should protrude above the section side channels, so that, if louvers are required, they can be easily installed. The same is true for induced-draft coolers. The plenum and section can be mated without a design change or interference problem.

Other features of a good design include headers large enough not to starve tubes while avoiding an unreasonable increase in pressure loss at the inlet or outlet. Tubes should be spaced far enough apart that, if a welded tube-to-tubesheet joint is required, it can be furnished without local welding heat damaging nearby fins. A good tube support design is necessary because it will hold tubes in position in service and during shipment; it will often influence tube spacing.

4.6 PLATE-FIN AND SPIRAL-FIN EXTENDED-SURFACE EXCHANGERS

It is logical to think of plate- and spiral-fin air coolers as competing with one another, since they are sized the same way. While true in theory, it seldom occurs in practice except in jacket water and lube oil cooler applications. Even in these services the units are seldom competitive in very small or very large units, but they are competitive in intermediate sizes.

In refinery and chemical plant applications, carbon steel is the preferred tube material. It costs less on a per pound basis than any metal. Users in these industries seldom specify steel tubes less than 1-in 12 BWG (0.108 in) thick. Standard tube and fin constructions include:

1. Wrapped aluminum edge wound fins on steel tubes
2. Extruded aluminum fins on steel tubes
3. Wrapping of an aluminum or steel fin in a spiral grooved-steel tube or welding of steel fins to steel tubes.

Constructions 1 and 3 above can employ aluminum-footed fins.

Aside from grooving, most of these methods of construction are independent of tube thickness. Coils needed in heavy industrial applications are usually large. Steel tube thickness can be increased for a small increase in price. Spiral fins on long tubes are finned efficiently. During manufacture the finning of short tubes is interrupted frequently to secure fin ends to prevent their unraveling. These interruptions reduce efficiency and add to cost. Spiral-fin manufacturers do not fin short tubes as economically as longer ones. Further, thicker tubesheets are used in most refinery, chemical plant, and utility applications (or their equivalent) than are associated with plate-fin units which are seldom thicker than 3/4 in.

Compare this with plate-fin construction commonly used in industrial applications. Most plate exchangers have thin tubesheets with tubes no thicker than 18 BWG (0.049 in). Tube materials are mainly aluminum, copper, or copper-based alloys, although tubes can be of any material. Tubes are expanded to contact the fins. Fin unraveling is not a problem. Plate-fin construction is generally not acceptable in refinery and chemical plants, for the following reasons:

1. The fin and tube combination has an operating temperature limit of about 200°F. Most process coolers operate above this temperature.

2. The presence of copper in any form is not acceptable if sulfur dioxide is present. In the event of a fire, copper loses strength at lower temperatures than steel, potentially allowing a product to feed a fire.

3. Tube gauges are thin by process standards.

4. Plate-fin tolerances are difficult to maintain. Fins from the same roll 2000 fins apart do not have the same tolerances. For this reason plate-fin bundles are difficult to assemble when tube lengths are 20 ft or longer. This is true for fins spaced 10 to the inch but not 4 or 5 to the inch. Process plants, in the main, use 30-ft 0-in or longer tubes without facing this problem.

5. The maximum thickness of steel tube that can be expanded economically the full finned length in plate-fin units is 16 BWG average wall. Thicker tubes could be used, but manufacturing costs would increase rapidly. Copper tubes can be expanded about two tube gauges thicker than steel because their tensile strength is less. By comparison, spiral fins are regularly applied to tubes 48 to 60 ft in length. In general, this process is independent of tube thickness, but not always. Thicker tubes are a necessity in process work.

6. Tubes installed in plate exchangers are generally not replaceable, while spiral-finned tubes are or can be.

7. Plate-fin units often have steel tubesheets. When this construction is used tube end connections are rolled joints without grooves or possibly one groove. Tubesheets and headers are nearly always thinner than 3/4 in. By contrast, this is the usual minimum thickness in spiral fin units. Plate-fin units often have copper or copper-base tubes and brazed U-bend connections. This construction is not acceptable in process work, because of the low melting point of these metals (possibility of fire) and loss of product (economics). Nearly all process coolers have single- or double-grooved rolled tube-to-tubesheet joints or welded tube-to-tubesheet connections.

8. Process units nearly always require a code stamp, while plate-fin units often do not. Plate fins are advantageous in some services. They are reliable and available at low cost. Spiral-fin units with short tube lengths are not as economical as plate-fin units. Exchangers with copper or copper-base tubes are ideal in water service, which is why jacket water coolers are a market mainstay. In general, the other market mainstay, lube oil coolers, has thin-gauge steel tubes and plate fins.

9. Plate-fin construction is, in the main, acceptable for jacket water service in refineries, chemical plants, and power plants because a leak to ground is seldom a hazard. They may not be acceptable for other reasons.

10. Cover-plate headers are common in plate- and spiral-fin units. Low design pressures allow for the use of thinner heads and covers. This construction is used in viscous and fouling services because it offers ease of cleaning. Plate fins do not have a bright future in process plants but are ideal in industrial applications where either plate or spiral fins can be used.

4.7 DUCT COOLERS

Duct coolers are another type of air-cooled heat exchanger. This type of coil is discussed in Section 6.1, under "Sections Installed in Ducts."

4.8 SELECTING SEVERAL AIR COOLERS TO DISSIPATE HEAT

One large air cooler may not be a practical solution for a given application even though it should be the least expensive. Examples would include discharging heat from a mechanical room through space-limited outer walls or discharging heat through the sides of mobile units. The space needed for a larger unit may block passageways, create structural interference problems, or be more useful for other purposes. Typically, smaller areas for discharging heat are available at other locations; the question is how best to use this space economically. Consider the case of several engines operating in parallel. The heat from these is to be dissipated using a series of air coolers all mounted on outside wall(s) and joined as one system. The number of coils is at the discretion of the manufacturer. A spare unit is to be provided and connected so that it is available when needed.

One way of approaching this selection follows. The method is useful if the choice is from a manufacturer's standard product line because the cost of the exchangers will be known from published prices. This method can be used in selecting other coolers, though their costs may not be as readily available. Begin by selecting one large unit to cool all engines. As an example, assume the exchanger selected has a face area of 12 ft by 20 ft, or 240 ft^2. This exchanger, or one with slightly fewer square feet of face area, will be near the required total face area for all engine coolers regardless of whether the choice is a few exchangers with large face areas or many smaller exchangers with lesser face areas.

Another way of expressing this is to divide the needed 240 ft^2 exchanger face area into any of the following:

- 2 exchangers each having 120 ft^2 of face area
- 3 exchangers each having 80 ft^2
- 4 exchangers each having 60 ft^2
- 10 exchangers each having 24 ft^2

The conclusion assumes that the same finned tube and spacing are used and the number of layers remains the same. More air per square foot of face area can be provided across smaller exchangers because more fan motor sizes are available. More air improves the MTD and results in somewhat smaller total face area than if fewer units with larger fans were selected. Fan horsepowers are easier to increase from $^1/_2$ to $^3/_4$ or $^3/_4$ to 1 hp than from 30 to 40 hp, because of drive considerations, though each is a 33% to 50% increase in energy.

Here are facts to consider when making a selection:

- Each additional unit requires another piping hookup.
- Make sure the wall height is enough to accommodate the exchangers selected.
- The cost of a spare unit increases as a lesser number of coolers are chosen. (Note: If only one unit were chosen, the spare unit would be another 12-ft by 20-ft unit.)

It should be understood that air-side face area and flow resistance are the same if the geometry is 12 ft by 20 ft or 20 ft by 12 ft or any combination that adds up to 240 ft^2. Square shapes are preferred for best fan coverage. The close-coupled design is preferred because it occupies less depth, leaving more space in aisles near the wall.

Here are possible choices to satisfy this example.

Number of coils	Cooler size	Considerations and recommendations
2 + 1 spare	(3) 11 ft by 11 ft	Space for this selection may not be available in the wall
3 + 1 spare	(4) 9 ft by 9 ft	Wall space may not be available. Nine feet is an unusual section width.
4 + 1 spare	(5) 8 ft by 8 ft	Probably a slightly smaller unit than this is required, but it may be ideal.
5 + 1 spare	(6) 7 ft by 7 ft	An unusual length and width section.
6 + 1 spare	(7) 6 ft by 6 ft	The best choice thus far.
7 + 1 spare	(8) 5 ft by 6 ft	Determine if a manufacturer's standard unit exists close to this in size, in which case this, too, may be a good choice.

The pattern to follow is as described above. Each must be checked to confirm that it meets the thermal requirements and that space is available for its use. A determination should be made about how the units are to be connected, with a rough layout made showing how this is done.

4.9 CALCULATING SURFACE AND AIRFLOW AREA FOR FIN COILS

The rating engineer may face at least three types of problems where standard construction does not provide the desired answer. These are:

1. Non-standard tube/fin construction may be the only realistic answer for a special metallurgy condition, when coolers must be of minimum weight or volume, when operating temperatures are high, or when the fin side gas is other than air.
2. Often the engineer must determine the capability of an existing cooler, built by others, and not of your standard construction. No performance data is available.
3. The guaranteed performance is not always correct and may be the reason why your unit is not performing. If the coil's surface and flow area are known an estimate of its cooling capacity can be reached.

In addition, problems may surface during the rating process that are best met using non-standard construction. Examples are:

- High flow and a higher than allowed pressure loss is the governing condition. Using a non-standard and larger tube diameter and fin will allow this need to be met.
- Small flows can be accommodated using a smaller diameter tube and fins.
- U-tubes of a specific diameter and with several passes may be the best choice to reduce the temperature drop per pass as these can cause thermal expansion cracking at tube bends. Several selections may be needed to determine the best choice for this application.

For these cases the engineer must provide a solution that meets these needs. Tube side conditions and the cooler space available will almost always be known. The unknowns are the coil's air side rate and pressure drop, total surface, and air flow area. The goal is to determine these. This frequently results in the use of non-standard tubes and fins or both. It is under these conditions that knowing a coil's surface and air side properties is essential. This effort is to establish the geometry and properties of non-standard coolers or the capacity of an existing cooler.

These surface options are available:

- Tube OD for both spiral and plate fin (1/4, 5/16, 3/8, 1/2, 5/8, 3/4, 7/8, 1, 1-1/4 and higher).
- Spiral fin height (1/8, 3/16, 1/4, 5/16, 1/2, 5/8, 9/16, 3/4 or others). When making a calculation, use the actual fin height, not the nominal ones listed.
- Plate fin height. Tube spacing may be the same as for spiral fins, but if needed, plate fin tubes can be spaced further apart. The question becomes, On what spacing should the tubes be? This is usually a pressure drop consideration. Dies for punching fin holes are expensive and, once built, usually cannot be changed to another size and spacing.

- Fins per inch (the usual limits are 4 to 14 for spiral fin, 4 to 20 for plate fin).
- Pitch (Usually triangular for spiral fin; triangular or rectangular for plate fin).
- Fin thickness (plate fins of copper or aluminum are usually thinner because, in manufacturing spiral-fins, thicker materials are needed to account the thinning that occurs in bending fins).
- Materials. Check with the manufacturers as many choices are available particularly for spiral fins.

An understanding of geometry, surface and coil flow area is beneficial when trying to determine why an existing exchanger is not performing, The problem needs correction or the exchanger adapted to new conditions. It is rarely a good idea to discard a workable exchanger. This problem is common for new and older units. The engineer seldom knows of the cooler existence until the problem arises. Its manufacturer may be out-of-business and drawings or operating data are unavailable. The details of the cooler will be unknown unless access to the exchanger is granted.

Coolers are usually sealed; to obtain the information needed requires the system be shut down and cooler head(s) removed, usually a plant problem. Once the details are known (tube diameter, length, pitch, fin number, height, spacing, material, number of tubes, passes, layers), the cooler's surface can be determined. Since the coil's performance is the goal, its tube spacing and surface will be compared with that of known coils to determine the approximate air side conditions. Tube side conditions are usually known.

Cooler manufacturers do not have calculated data for all tube/fin combinations. They have these for many including their outside to inside surface ratios. Tubes placed in an air flow path obstruct flow, increase pressure loss, and effect the selection of fans or blowers (coverage, noise, blade angle, and aspect ratio). The usual way to obtain information about spiral fin coils is given in Table 4-3, "Calculation of Spiral-Fin Coil Surface," and in Table 4-4, "Calculation of Spiral-Fin Coil Airflow Area." Remember, a selection is not complete until fan coverage and/or the duct aspect ratio needs are met.

Knowing how to calculate external surface is useful in establishing performance. Fin and bare tube surfaces that are in contact with cooling air are considered as external surface. The following describes the external surface areas of spiral fin tubes (See Figure 4-6).

1. Heat can enter at the tips of fins. These are the cylindrical areas having a diameter equal to the fin OD. The thickness is that of the original fin. This is not strictly speaking correct because the fin is thinner due to the metal stretching that occurs during the bending process. The error introduced by thinning is usually small. Don't overlook the number of fins per foot in the calculation.
2. Heat passes through both areas of the fin's doughnut shaped sides. Its larger diameter is the OD of the fin and the smaller is the OD of the tube. Include the number of fins per foot in the calculations.

Table 4-3 Calculation of Spiral-Fin Coil Surface

Surface area of fin tops, ft^2/ft of tube	$(\pi DtnN)/144$
Surface area of fin sides, ft^2/ft of tube	$(\pi/4)(D^2 - d^2)(2nN)/144$
Surface area of fin covering tube, ft^2/ft of tube	$(-)\pi(dtnN)/144$
Surface area of bare tube, ft^2/ft of tube	$(\pi dN)/144$
	Total ft^2/ft of tube

where
- D = outside diameter of fins, in
- d = inside diameter of fins or outside diameter of tube, in
- n = number of fins per inch
- N = number of inches per foot
- 2 = number of fin sides
- h = fin height, in
- t = fin thickness,* in

Example: 1-in-OD tube (11 BWG), ID = 0.76 in; 2-1/4-in-OD fins, (11) 5/8-in fins per inch, 0.012 in thick:*

Area of fin tops = (3.1415)(2.25)(0.012)(11)(12)/144 =	0.0778
Area of fin sides = (3.1415/4)(2.25^2 -1^2)(2)(11)(12)/144 =	5.849
Area of base of fins = (3.1415)(1)(0.012)(11)(12)/144 =	(–)0.0346
Area of bare tube = (3.1415)(1)(12)/144 =	0.262
Surface area (ft^2/ft of tube) =	6.154

Inside tube surface = (3.1415)(0.76 in ID)(12 in/ft)/144 = 0.200 ft^2/ft of tube
Outside/inside surface ratio = 6.154/0.200 = 30.83

A coil with 285 tubes 30 ft long (effective fin length 29.5 ft) would have a surface area of (6.154)(285)(29.5) = 51,739 ft^2.

*Each fin is thinner at its tip.

3. Heat is transferred through a tube's bare tube surface, that is, through the areas not covered by fins. This area is calculated by using the tube's bare surface and deducting that part of the surface covered by fins.
4. Heat will not be directly transferred through the bare tube areas covered by fins because there is no direct air/surface contact. This area is calculated the same way as the fin tip area (point 1) except that the smaller fin diameter is used and its value is subtracted from the bare tube area.

Using these calculations one can determine the surface per foot for spiral fin tubes, the outside to inside surface ratio, and the exchanger's surface. The method is outlined in Table 4-3 and Figure 4-6 with a sample calculation included.

The air flow area is the gross face area of the cooler less the space occupied by the tubes; in other words, it is the area through which air passes. The method of calculating a unit's air flow area is given in Table 4-4 Calculation

Table 4-4 Calculation of Spiral-Fin Coil Airflow Area

For same tube and spacing as in Table 4-3.

Obstruction of tube (per foot of length) = 1 in × 12 in	12.0 in^2
Obstruction of fins = $thnN(2)$ = (0.016)(5/8)(11)(12)(2)*	1.98 in^2
Obstruction per foot of tube	13.98 in^2

where

t = fin thickness, in
h = fin height, in
n = number of fins per inch
N = number of inches per foot
2 = number of fins on each side of tube center line

Consider a coil with 48 tubes in a layer spaced 2-5/8 in apart which includes a clearance at the seal strips. The width of the coil for airflow is:

$$(48)(2\text{-}5/8 \text{ in}) = 126 \text{ in}$$

The tube length is 30 ft. From this deduct 2 in for tube-to-tubesheet joints and also the surface at 5 tube supports each 1 in wide, or a total length of 360 in -7 in = 353 in of length for airflow.

Gross area for airflow = (126)(353) =	44,478 in^2
Less fin-tube obstruction = (13.98)(29.416)(48) =	19,739 in^2
Area for airflow across coil	24,739 in^2
	or 55.6% of the gross area for air flow

To obtain the overall section width, the following must be accounted for:

Width of the header end plates
Space for header movement
Effect of seal strips and clearances
Thickness of the side channels

*For resistance, one part of the fin is above the tube centerline and the other is below the centerline.

of Spiral-fin Coil Airflow Area and in Figure 4-6 The Geometry of the Cooler in Tables 4-3 and 4-4. As a general rule approximately 1 in of space is needed for structural considerations on the cooler sides. In the example, Table 4-4, the area for air flow was 55.6% of the gross area, a relatively high percentage because the tubes are spaced far apart. For estimating purposes only, this value is rounded off to 50% for 1-in tubes with high fins ($^1/_2$ in or $^5/_8$ in), a handy number to have when a quick estimate is needed regarding the feasibility of a cooler in terms of space.

Tubes in spiral and plate fin exchangers are usually spaced uniformly. Some installed spiral-fin tubes have their fins touching one another; in others there is up to 5/16 in clearance between fin OD and side containing member. This distance should be minimal to avoid bypassing.

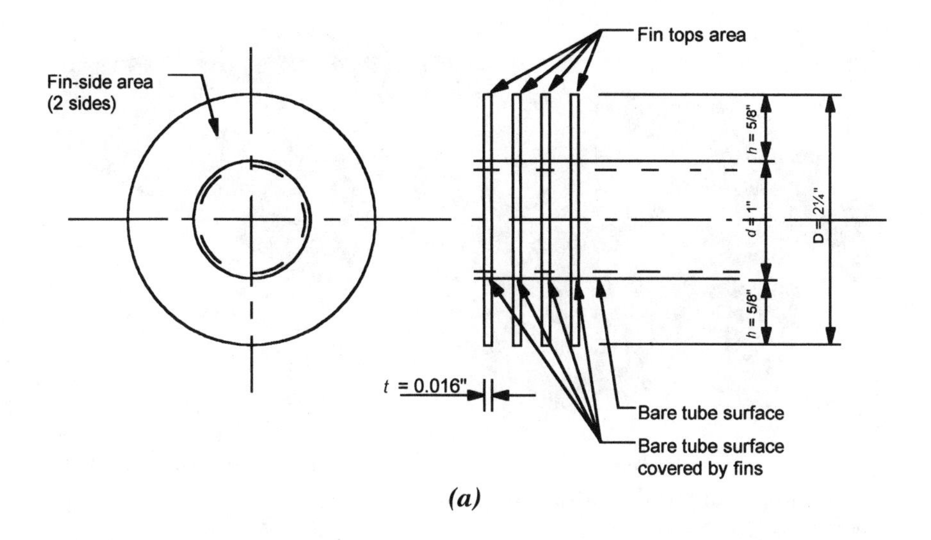

(a)

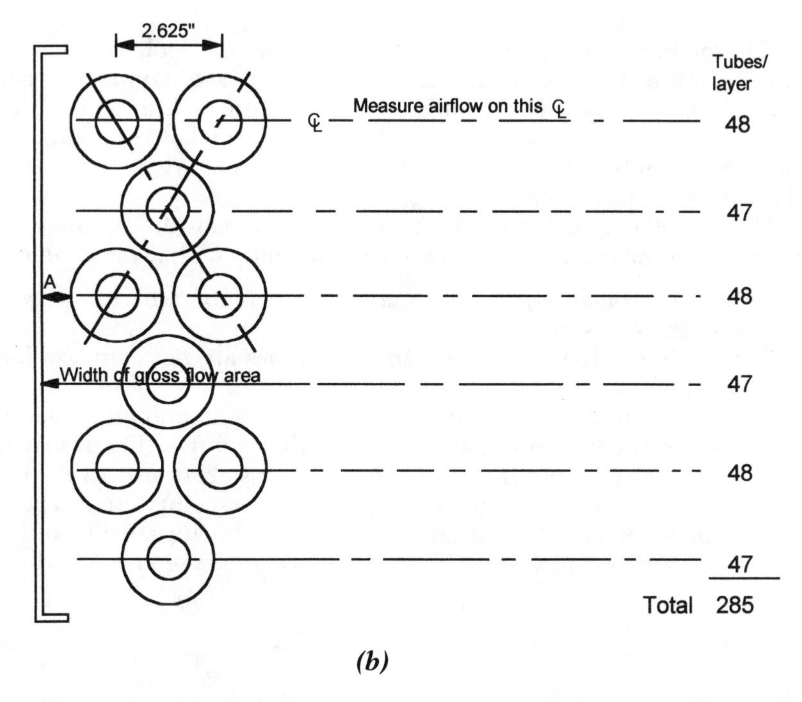

(b)

Figure 4-6. Cooler geometry used for Tables 4-3 and 4-4. (*a*) Spiral-finned tube surface. (*b*) Spiral-finned tube airflow area. A = clearance between side member or seal strip and fin OD.

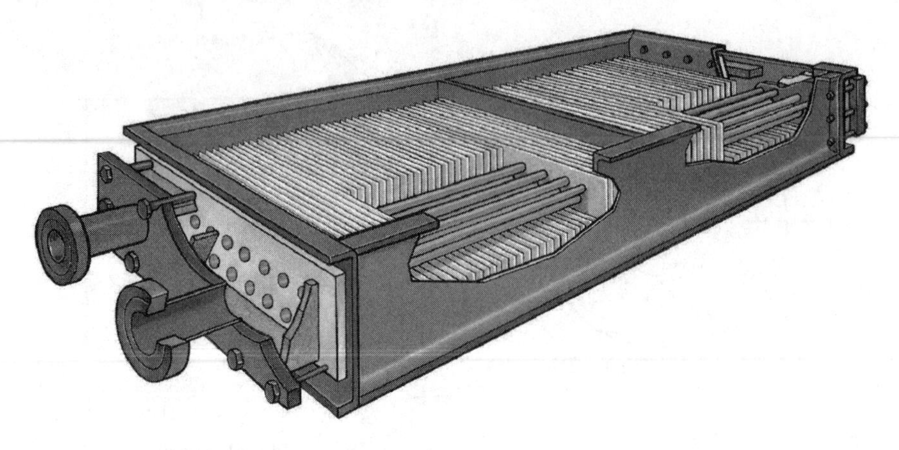

Figure 4-7. Motor or generator air cooler. (Reprinted by permission of BDT Engineering, Industrial Products Division.)

The amount of surface in a plate fin exchanger is calculated the same way as for spiral fins. The difference is that plate fins are stamped for banks of tubes; hence, surface is calculated for banks of tubes. Either way, the result is the same. External surface includes the area of both fin sides, fin tips (outer edges only), and bare tubes less the bare tube area covered by fins. A plate fin exchanger is shown in Figure 4-7.

The calculated surface and air flow area of plate fin construction are needed as often as for spiral fins for size feasibility with these limitations.

1. Plate-fin construction is generally not acceptable where large temperature gradients exist.
2. Large flows requiring large diameter tubes are not common for plate fin units.
3. Plate fins units are usually made of copper or aluminum; other materials are rarely used unless they are of thin lightweight construction. It takes considerable energy to expand thick steel or other tube material to contact fins, enough to make them non-competitive. The use of stronger materials is limited due to manufacturing difficulties. The operating pressure at which these units can be used is limited.

Chapter 5

Extended-Surface Metallurgy

Exchangers can be less expensive and smaller if finned tubes are used. Fins make better use of the pressure drop available, because the length of travel in pure crossflow is low and the available flow area is large. This is advantageous in low-pressure gas or vapor applications where the pressure drop available is small. Finned tubes have been used in air coolers, shell-and-tube units, and longitudinal pipe exchangers. This chapter presents some facts about them.

5.1 SELECTING ALUMINUM OR COPPER FINS

A question often asked is, Should the fin material in spiral or plate fin units be aluminum or copper? Copper is the better conductor, so the follow-up question is, How much better? The answer depends on whether the coolant is air or water. For air cooling, Table 5-1 was developed and illustrates the effect that each material has on the overall heat transfer rate, U. Three examples, each operating at the same air side conditions are given.

The tabulation in Table 5-1 for spiral fins shows that when air is the coolant, the use of copper fins nearly always results in exchangers that are 2 to 4% smaller than those that use aluminum fins. Historically, copper prices fluctuate more than aluminum and cost more on a per pound basis, usually 1.25 to 2.5 times more. The specific gravity of copper is 3.25 times greater than aluminum. While copper may be 10% thinner than aluminum, this is not to say that copper fins will be thinner than aluminum fins although they often are. These factors are additive and explain why, together, copper fins cost 4 to 7 times more than aluminum on a per pound basis. The differential applies to both spiral and plate fins and is the main reason why aluminum fins are used more often than copper. Either could be used on most applications.

Other factors affect the choice of fin material. Copper is best where vibration is a problem. Pieces do not break off from copper as readily as from aluminum as a result of vibration. In generator cooler service this is a plus, because it reduces the potential for winding damage if pieces of metal flake off. Copper is sturdier at higher temperature and is usually best in atmospheres containing chemicals.

The overall heat transfer rate using copper or aluminum longitudinal fins on steel tubes in liquid service is roughly the same using either material. The most popular construction is steel fins welded to steel tubes. One way of bonding non ferrous fins to steel, or to most other tube materials is to insert fins into plowed grooves and then force the displaced metal

Table 5-1 Variation in Air Cooler Overall Heat Transfer Rate U Using Aluminum or Copper Spiral Fins

Three examples, each with a different tube side resistance, show that copper fins are 2-1/3 to 3% more efficient than aluminum fins 1-in-OD steel tubes with eleven 5/8-in-high fins per inch

	Spiral fin material					
	Example 1		Example 2		Example 3	
	Aluminum	**Copper**	**Aluminum**	**Copper**	**Aluminum**	**Copper**
Shell film resistance*	0.096	0.096	0.096	0.096	0.096	0.096
Shell-side fouling resistance*	0.000	0.000	0.000	0.000	0.000	0.000
Metal resistance*	0.0189	0.0144	0.0189	0.0144	0.0189	0.0144
Tube film resistance*	0.0201	0.0201	0.0201	0.0201	0.0201	0.0201
Tube fouling resistance:*						
0.001	0.0201	0.0201	—	—	—	—
0.002	—	—	0.0402	0.0402	—	—
0.003	—	—	—	—	0.0603	0.0603
Total resistance R*	0.1551	0.1506	0.1752	0.1707	0.1953	0.1908
U**	6.4475	6.640	5.708	5.8582	5.120	5.241
% variation in U	2.99		2.63		2.36	

*Resistance R in units of $h \cdot ft^2 \cdot °F/Btu$.
**Heat transfer rate U in units of $Btu/(h \cdot ft^2 \cdot °F)$.

against the fin, thus creating the bond. The overall heat transfer rate, U, using this construction, is approximately 20 to 35% better than if steel tubes and steel fins were used. There are times when the heat transfer rate improvement can be better. This occurs when the number of units in series is reduced by one. For example, a 12 psi allowable pressure drop over four units in series averages 3 psi pressure drop per unit. If only three units are required, the pressure drop average per unit is 4 psi. This increase in pressure drop per unit plus the the effect of better thermal conductivity fins substantially improves the overall heat transfer rate, U. Few companies, however, now produce plowed-groove units mainly because most fins to tubes connections are made by welding, and non-ferrous materials cannot be welded to steel tubes.

5.2 FINS OF MATERIALS OTHER THAN COPPER OR ALUMINUM

Most metals other than aluminum and copper are relatively poor conductors of heat (see Table 5-2). However, some materials, particularly steel, are less expensive on a per pound basis. The question is, Why not use this metal more often as the fin material?

An advantage of aluminum and copper is that they do not require a large amount of energy to be formed into fins. Other materials including steel do. Nonferrous fins can be relatively thin compared with fins of other metals. Copper and aluminum fins in plate-fin units are regularly in the 0.007-in-thick range. Wrap-on fins must be made of thicker material to prevent tearing of metal during the finning process. The starting thickness of material before bending copper or aluminum into fins is around 0.016 in for $^5/_8$-in-high fins wound onto 1-in-OD tubes. Copper fins can be about 10% thinner than aluminum before manufacture begins. Most spiral steel fins, on the other hand, are at least 50% thicker than aluminum fins when manufacturing begins, meaning that fewer fins per inch can be wrapped on a tube. The usual limit for copper or aluminum fins is 12 fins per inch on 1-in-OD tubes, while for steel it is 9.

The reasons spiral steel fins are rarely used in air coolers can be summed up as follows:

1. The maximum extended steel surface (spiral fin) that can be added to a foot of tubing, while it varies with fin thickness, is roughly 60 to 75% that of aluminum or copper. More sections are needed for this reason.
2. Steel fins must be thicker to prevent tearing the fins during manufacture.
3. More energy is needed to bend steel because of its greater strength and thickness.
4. Steel costs about $^1/_3$ as much as aluminum on a per pound basis, but its density is 3 times as great (and prices vary daily). Hence, savings using steel fins are minimal.
5. The metal resistance of steel fins is higher. More sections are needed for a comparable exchanger in the same service.

Table 5-2 Thermal Conductivity of Metals, Btu/(h · ft · °F)

	Temperature, °F															
Material	**70**	**100**	**200**	**300**	**400**	**500**	**600**	**700**	**800**	**900**	**1000**	**1100**	**1200**	**1300**	**1400**	**1500**
Carbon steel	30.0	29.9	29.2	28.4	27.6	26.6	25.6	24.6	23.5	22.5	21.4	20.2	19.0	17.6	16.2	15.6
C–1/2 moly steel	24.8	25.0	25.2	25.1	24.8	24.3	23.7	23.0	22.2	21.4	20.4	19.5	18.4	16.7	15.3	15.0
1 Cr–1/2 Mo & 1-1/4Cr–1/2 Mo	21.3	21.5	21.9	22.0	21.9	21.7	21.3	20.8	20.2	19.7	19.1	18.5	17.7	16.5	15.0	14.8
2-1/4Cr–1 Mo	20.9	21.0	21.3	21.5	21.5	21.4	21.1	20.7	20.2	19.7	19.1	18.5	18.0	17.2	15.6	15.3
5 Cr–1/2Mo	16.9	17.3	18.1	18.7	19.1	19.2	19.2	19.0	18.7	18.4	18.0	17.6	17.1	16.6	16.0	15.8
7 Cr–1/2 Mo	14.1	14.4	15.3	16.0	16.5	16.9	17.1	17.2	17.3	17.2	17.1	16.8	16.6	16.2	15.6	15.5
9 Cr–1 Mo	12.8	13.1	14.0	14.7	15.2	15.6	15.9	16.0	16.1	16.1	16.1	16.0	15.8	15.6	15.2	15.0
3-1/2 nickel	22.9	23.2	23.8	24.1	23.9	23.4	22.9	22.3	21.6	20.9	20.1	19.2	18.2	16.9	15.5	15.3
12 Cr & 13 Cr	15.2	15.3	15.5	15.6	15.8	15.8	15.9	15.9	15.9	15.9	15.8	15.6	15.3	15.1	15.0	15.1
15 Cr	14.2	14.2	14.4	14.5	14.6	14.7	14.7	14.8	14.8	14.8	14.8	14.8	14.8	14.8	14.8	14.8
17 Cr	12.6	12.7	12.8	13.0	13.1	13.2	13.3	13.4	13.5	13.6	13.7	13.8	13.9	14.1	14.3	14.5
17–19 Cr (TP 439)			14.0													
TP 304 stainless steel	8.6	8.7	9.3	9.8	10.4	10.9	11.3	11.8	12.2	12.7	13.2	13.6	14.0	14.5	14.9	15.3
TP 316 & 317 stainless steel	7.7	7.9	8.4	9.0	9.5	10.0	10.5	11.0	11.5	12.0	12.4	12.9	13.3	13.8	14.2	14.6
TP 321 & 347 stainless steel	8.1	8.4	8.8	9.4	9.9	10.4	10.9	11.4	11.9	12.3	12.8	13.3	13.7	14.1	14.6	15.0
TP 310 stainless steel	7.3	7.5	8.0	8.6	9.1	9.6	10.1	10.6	11.1	11.6	12.1	12.6	13.1	13.6	14.1	14.5
2205 (S31803)	8.0	8.5	9.0	9.5	10.0	10.5	11.0	11.5	12.0							
3RE60 (S31500)	8.4	8.5	9.0	9.4	9.8	10.2	10.6	11.0	11.3							
Nickel 200			38.8	37.2	35.4	34.1	32.5	31.8	32.5	33.1	33.8					
Ni–Cu (N04400)	12.6	12.9	13.9	15.0	16.1	17.0	17.9	18.9	19.8	20.9	22.0					
Ni–Cr–Fe (N06600)	8.6	8.7	9.1	9.6	10.1	10.6	11.1	11.6	12.1	12.6	13.2	13.8	14.3	14.9	15.5	16.0
Ni–Fe–Cr (N08800)	6.7	6.8	7.4	8.0	8.6	9.1	9.6	10.1	10.6	11.1	11.6	12.1	12.7	13.2	13.8	14.5
Ni–Fe–Cr–Mo–Cu (N08825)			7.1	7.6	8.1	8.6	9.1	9.6	10.0	10.4	10.9	11.4	11.8	12.4	12.9	13.6
Ni–Mo alloy B		6.1	6.4	6.7	7.0	7.4	7.7	8.2	8.7	9.3	10.0	10.7				
Ni–Mo–Cr alloy C-276 (N10276)		5.9	6.4	7.0	7.5	8.1	8.7	9.2	9.8	10.4	11.0	11.5	12.1			

Aluminum alloy 3003	102.3	102.8	104.2	105.2	106.1											
Aluminum alloy 6061	96.1	96.9	99.0	100.6	101.9											
Titanium (grades 1, 2, 3, 7)	12.7	12.5	12.0	11.7	11.5	11.3	11.2	11.2	11.2	11.3	11.4	11.6				
Admiralty			70.0	75.0	79.0	84.0	89.0									
Naval brass			71.0	74.0	77.0	80.0	83.0									
Copper			225.0	225.0	224.0	224.0	223.0									
90–10 Cu–Ni			30.0	31.0	34.0	37.0	42.0	47.0	49.0	51.0	53.0					
70–30 Cu–Ni (C71500)			18.0	19.0	21.0	23.0	25.0	27.0	30.0	33.0	37.0					
7 Mo (S32900)		8.8	9.3	9.8	10.3	10.8	11.3									
7 Mo plus (S39250)		8.6	9.4	10.2	11.1	11.8	12.7									
Muntz			71.0													
Zirconium			12.0													
Cr–Mo alloy XM-27			11.3													
Cr–Ni–Fe–Mo–Cu–Cb (alloy 20CB)			7.6													
Ni–Cr–Mo–Cb (alloy 625)	5.7	5.8	6.2	6.8	7.2	7.7	8.2	8.6	9.1	9.6	10.1	10.6	11.0	11.5	12.0	12.6
Al 29–4–2	8.8															
Sea-Cure	9.4	9.6	10.3	10.9	11.6	12.3	12.9	13.7								
Al-6XN (N08367)			7.9													

References:
ASME Section II, 1998 edition
Huntington Alloy, Inc., Bulletin No. 15M1-76T-42
AIME Tech. Pubs. 291, 360, and 648
Allegheny Ludlum Steel Corp.
Teledyne Wah Chang Albany
Trans. ASST 21: 1061–1078
Babcock & Wilcox Co.
American Brass Co.
Trent Tube
Airco, Inc.
Cabot-Stellite
Carpenter Technology
International Nickel Co.
Sandvik Tube
Source: Reprinted by permission of the Tubular Exchanger Manufacturers Association (TEMA)

Steel fins and those of higher-quality alloys are ideal in many applications. Steel tubes and fins are used as oil tank heaters because their fin temperature is nearer the temperature of the heating steam than that of the oil, and this is usually above the operating temperature of aluminum. Using steel fins simplifies the cleaning of tanks, because steel fins are not easily damaged in service and they can withstand caustic cleaning.

At high tip temperatures, the fin tips can burn. This problem can be avoided in either of two ways. One is to make the fin tip thicker, which can be achieved by starting the finning process with a thicker fin material. The other is to upgrade the fin material using an alloy. This problem commonly occurs in high-temperature applications such as furnaces.

Note that most metals listed in Table 5-2 have lower thermal conductivity values than steel. This means that more or thicker fins will be needed for a given application if these materials are used in place of steel.

5.3 LOWFIN TUBES

Lowfin tubes can be made of copper, admiralty, red brass, aluminum brass, aluminum bronze, the copper nickels, stainless steels, and carbon steels. Check with manufacturers about your specific need, as many other metals can be finned.

In terms of metallurgy, consider these features of lowfins:

1. Fewer tubes are needed using lowfins than bare tubes. The lesser quantity of material needed may result in a lower-cost unit.
2. The tube and fin are integral, so there is no chance for a fin to unravel.
3. Lowfin tubes have smaller inside diameters than a comparably sized bare tube. This may result in a shorter-length unit overall (refer to Section 2.4).
4. Higher fin counts, 28 to 32 per inch, can be built and are beneficial where the fin-side allowable pressure is relatively large as in high-pressure steam service.

For other facts about lowfins, see Sections 14.2 and 17.7.

5.4 SOME RULES OF THUMB FOR FINNED SURFACES

Here are general rules to follow when selecting a finned surface; they are observations noted by the writer over a career and are influenced as much by economic as by heat transfer considerations:

1. Integral fin construction is usually preferred when a liquid is on the finned side.
2. The greater the ratio of fin-side rate to tube-side rate, the greater the OA/IA ratio should be.
3. The ratio of the outside area of a finned tube to the inside area of the bare tube can be reduced as the outside film rate approaches the tube-side rate. In other words, fin height or number of fins per inch can be reduced.

4. As the operating pressure of the fin-side fluid increases, the needed flow area will be less. This change can be made by:
   a. Decreasing the spacing between tubes. If this option is chosen, make sure a tube support design is available, particularly for air coolers.
   b. Reducing the number of tubes per layer and increasing the number of layers crossed.
   c. Either or both of the foregoing in conjunction with shortening the tube length.
5. The fin-side rate will improve as the operating pressure increases, usually because more pressure drop is available. Note: As a comparison, an allowable pressure loss of 3 psi is 83 times more than the 0.5 in usually allowed for air. This results in a substantial rate improvement, a big advantage in superheated-steam applications. Lowfins can have up to about 30 fins per inch (the maximum number depends on the strength of the base material). A high fin count is beneficial in superheated-steam service.
6. Most finned surfaces have lower operating temperature limits than bare tubes.
7. Tubes with wrap-on ferrous fins have a lower surface per foot of tube length than those using copper or aluminum fins because thicker steel is needed to successfully stretch the material without tearing the fin metal. It means there will be fewer fins per inch and a higher metal resistance because of the poorer thermal conductivity of steel.
8. For tubes of the same material, diameter, and thickness also having fins of the same diameter, material, and thickness and the same surface per foot, the metal resistance is the same whether longitudinal or spiral fins are used, provided that their outside heat transfer rates are the same.

Chapter 6

Air Cooler Fans or Blowers

Fans or blowers are usually necessary for air coolers to perform. Their selection has a pronounced effect in deciding which air cooler to install. Some variables that might influence selection include:

Fan coverage and bundle geometry
Fan tip speed and noise
Whether to use fan(s) or blower(s)
Fan or blower starvation
Cost
Air velocity limit
Geometry of transition pieces connecting blower and coil or coil and duct
Effect of fan size on shipping width
Meeting cfm and static pressure requirements

In addition, the engineer should consider several question before finalizing a selection if trouble jobs are to be avoided: Should the fan blades be fixed or autovariable? Should the fan motor be single-speed or two-speed? Will the motor be overloaded during low-temperature operation? How many fans should be used? Should the design be forced-draft or induced-draft? And is recirculation needed (to prevent freezing or channeling)?

6.1 FAN COVERAGE AND BUNDLE GEOMETRY

The selection of an air-cooled unit is not final until the subject of fan coverage and bundle geometry is decided. Different rules apply depending on the application; these are best illustrated by example.

6.1.1 Small, Noncode, Close-Coupled Air-Cooled Units

Similar designs are commonly used in services having similar physical requirements. Jacket water coolers, glycol/water coolers, lube oil coolers, transmission oil coolers, and radiators usually use similar, small exchangers. Most such exchangers are mass-produced. In these services most coils have face areas of less than 5 ft by 5 ft, although some are as large as 10 ft by 10 ft. When a propeller fan and coil are assembled with less than about 6 inches' clearance between them, the construction is described as direct-connected. Units of this construction are often mounted in a vertical plane. Care must be taken to see that that horizontally discharged hot air will not be an irritant or a safety issue to personnel.

Consider an installation having an 8-ft-diameter fan direct-connected to a 10-ft by 10-ft coil. The fan area is 50.3 ft^2, the section area is 100 ft^2, and the fan coverage is 50.3%, which is unacceptably low for direct-connected units. A minimum fan coverage has to exist for air to be properly distributed over the coil. A fan 9 ft 6 in diameter (71% coverage) would be acceptable, but a fan nearer 10 ft in diameter would be better. Higher fan coverage ratios are needed when units are direct-connected so that air distribution will not be a problem.

6.1.2 Process Air Coolers

Forced-draft air coolers have a plenum height in the range of 3 to 5 ft or more depending on the coil face area. This space allows air to be evenly distributed across the bundle. It differs from the close-coupled unit described in Section 6.1.1. Plenum requirements are similar for forced- or induced-draft units. In both designs the requirement is that fan coverage be a minimum of 40% of the bundle face area. When two fans per bay are furnished, the requirement applies to one-half the coil face area per fan (Figure 6-1).

6.1.3 Sections Installed in Ducts

Air-handling units generally cool air in ducts with water or with a glycol/water solution in the tubes. The air mover is not a fan but a blower. A problem to be faced is that warm duct air must pass through a transition piece connecting the face area of the coil to the duct. The longer the transition piece, the lower the pressure drop, because the expansion of the airstream is less severe. After leaving the coil, the cooled air contracts, requiring a transition piece to connect the coil to the continuation duct. The trade-off in transition piece designs is that the dimensions must be reasonable (the available space and needed space are not always compatible). Transition pieces must provide good distribution across the coil. Trade-offs include:

1. The greater the aspect ratio, the thicker the duct metal, meaning higher cost plus more supports and hangers.
2. When dimensionally limited, the more severe the ratio of coil face area to duct face area, the shorter the transition piece and the higher the pressure drop through these connecting pieces.
3. In contrast, the longer the transition pieces, the lower the expansion and contraction losses. In retrofit situations, space is usually not available to make this choice.
4. The longer the coil, the less expensive the coil will be. Transition pieces cost more when the aspect ratio is high (usually over 3 to 1).

For these reasons and for good distribution, the following rules of thumb are good starting points to consider before finalizing the selection of coil and transition pieces (keep in mind that it is not always possible to stay within these bounds):

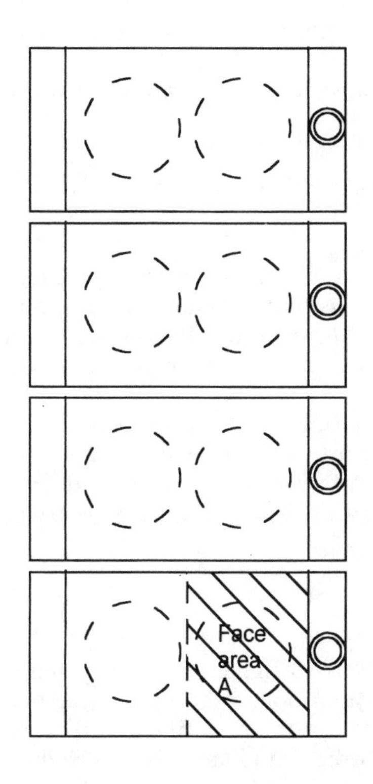

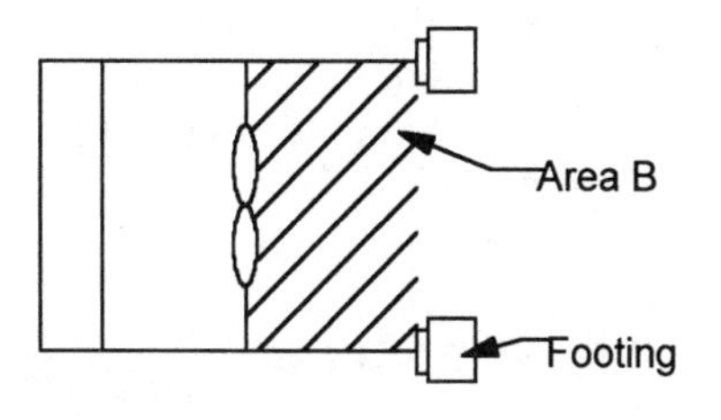

NOTES:

1. The fan area must be at least 40% of face area A.

2. Area B must be 25% or more larger than the fan area to avoid starving the fans in bays 2 and 3. Bays 1 and 4 also receive air through side openings.

3. Make all column footings the same size including those on the outside columns. These will be needed if there is a future expansion.

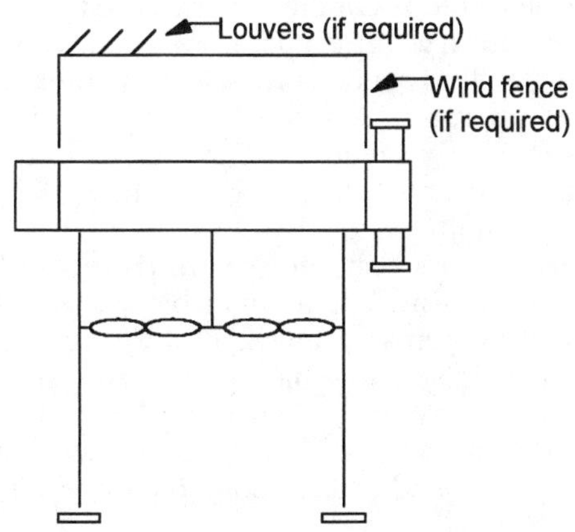

Figure 6-1. Typical four-bay ground-mounted forced-draft air cooler.

1. To keep the cost of coils and transition pieces low, the length-to-width ratio should be no more than 3. This ratio keeps the duct weight and cost within reasonable bounds. At times width will be as limiting a factor as length.
2. The combined angles of the transition piece should not total more than 30°.

6.2 FANS AND BLOWERS

An air mover is selected based on the cfm it must deliver and the pressure loss to overcome. Be sure to include with the coil loss that of heaters, louvers and any other pressure loss condition that applies. The heat transfer rate across an extended surface bundle depends on the air velocity through it. These conditions are independent of whether a fan or a blower is to be used. As a rule the choice of air mover will be obvious after considering the conditions that follow.

Using the same horsepower, air movers can be described in these terms. Fans deliver large volumes of air at low static pressures; blowers deliver lesser volumes of air at higher static pressures. The rating engineer will decide which is best for the application.

6.2.1 Type of System

Most air coolers, including A/C condensers, are designed as open systems, that is, the cooling air is continuously changed. Coolers should be designed and positioned to prevent warm exhaust air from recirculating and reintering the inlet air stream. Evaporators, on the other hand, are part of the duct system in the sense that the cool air delivered is contained; however, the system is open-end at the make-up air entrance. In evaporators air is cooled rather than heated and this may be enough to condense moisture on the coil. Delivering recirculated room air to evaporators is a plus because this air is already cooled and its reuse results in energy savings. A/C system recirculation can be a negative if some exhaust air enters the system at the makeup air inlet. This remnant will add to the system load.

Compare the quantity of air delivered to an air cooler with the quantity delivered to an evaporator, or said another way, Will a fan or blower be best for the application? The quantity of air that crosses an evaporator is limited, usually, 400 cfm per ton of cooling. This air is in a partially contained system. The air mover is not sized for the pressure loss across the coil but for the total pressure loss in the system. Included in this are the coil, length of travel, fittings, heaters, air filters, louvers, registers and grilles. Combined, the total drop will be in the 2 to 10 in range historically.

One may think that because more static pressure is available using a blower, more is also available for moving air across an evaporator. This is not true because a 500 ft/min velocity limit (Section 17.1) must be maintained to keep moisture from blowing onto ductwork. Higher velocities increase a system's pressure loss and may cause a change in the cooling capacity of a system. Further, higher duct velocities, above 1200 ft/min, in low-pressure ducts often generate unwanted air noise (Section 17.3). Remember, noise travels in both directions in a duct while air and vapors travel in one direction only. Retrofitting ducts with noise damping material is usually space limited, particularly at inlets to the system or in areas ahead of the blower. Because of this, the desired result is not as easy to obtain as one would like. The passage of air through a system having a high pressure loss (over about 0.75 in) cannot be economically satisfied by a propeller fan but is easily met using a blower. Consider a stand alone air cooler. Its cooling air passes over the coil but does not have to traverse ducting, fittings, filters, registers, grilles and often not through heaters or louvers. Its air flow is not limited by duct size, transition pieces or a cfm per ton limit. Thus, the rating engineer can select a flow knowing that a coil's typical pressure loss is in a range common to most fans (0.4 to 0.6 in). Should the air side pressure loss have to be lowered it can be done by spacing tubes on a wider pitch, reducing the number of layers (Section 14.8), or adjusting the quantity of air flowing. In general these conditions can be met by selecting the proper fan. Again the air mover type to recommend is obvious.

6.2.2 Geometry Considerations

Fans and blowers must meet other conditions. For fans the geometry limitation is coverage (Section 6.1); the comparable for ducts is meeting aspect ratio requirements (Section 6.1.3). Both must avoid conditions that starve the air mover (Section 20.2).

6.2.3 Noise

The air cooler noise level can be lowered by reducing the fan's tip speed and blade angle (Sections 4.3 and 17.3) while blower noise can be lowered by reducing speed. Duct noise is a companion problem that can generally be lowered by increasing duct size. In either case blower connections must be considered.

6.2.4 Other Factors

Consider the possibility of recirculation occurring for fans or blowers (Chapter 20). Also, both air mover types have operating temperature limits. A fan motor operating at low temperature sometimes becomes overloaded due to increasing air density. Air cooler drives may operate over temperature ranges of –60°F in winter to +150°F in summer while an evaporator blower may vary by no more than 25°F over a year.

6.3 FAN DRIVES

A fan drive has several components but the most important are the fan and driver, usually an electric motor. There may be more than one drive per bay.

This article centers on electric motor drives but the data also applies to hydraulic motors. A speed reducer is often needed; couplings, bearings and base plates nearly always are. The popular types of electric drives are 1) a fan directly connected to a motor, 2) a V-belt drive speed reducer located between the fan and driver, and 3) a gear reducer is used in place of a V-belt drive. Some designs and components are shown in Figure 4-2, Typical Drive Arrangements.

The principal variables in fan drives are motor rpm and fan diameter. Keep in mind that fan drives are seldom built by exchanger manufacturers; drives are part of the cooler system. Their selection must be economical if the cooler is to be economocal.

For 110 V, 3 phase, 60 cycle electric motors these standard rpm choices are available: 690, 860, 1160, 1750 and 3550. The propeller fan tip speed limit is usually 13,500 ft/min provided noise is not a problem and roughly 10,750 ft min if it is a problem.

Tip speed is calculated as follows.

$$\text{tip speed} = (3.1416)\ (\text{fan diameter})\ (\text{fan rpm})$$

Consider two separate installations. One has a 5 ft diameter fan and a motor rpm of 690. The other is a 14 ft diameter fan with a tip speed of 10,750 ft min. The desired data for each can be obtained using the formula for tip speed.

$$(3.1416)(5)(690) = 10{,}838 \text{ ft min for the 5 foot diameter fan}$$
$$(3.1416)(14)(\text{rpm}) = 10{,}750 \text{ ft min.}$$
$$\text{rpm} = 244 \text{ for a 14 ft diameter fan.}$$

Note in the example using a 5 ft diameter fan with a direct connected motor that the tip speed is only slightly above the value that normally avoids a noise problem. It is important that the lower rpm motor be chosen. By inserting other fan diameter values in the tip speed formula it can be seen that by varying the fan diameter and motor rpm a combination of these can be selected that satisfies the noise limit. The higher rpm motors are usually less expensive. For this reason fans less than 5 ft diameter are usually direct connected to the driver eliminating the need for a speed reducer.

A review of the performance curves of several large diameter fans (over 14 ft) shows that they are about the maximum diameter fan that can be operated using a 30 HP or less fan motor. Note the slow rpm of the 14 ft diameter fan given in the example. Above 30 HP most users prefer to use a reduction gear rather than a V-belt drive. Between the extremes of these two examples the speed reducer is usually a V-belt drive. This applies whether the the cooler has a spiral or plate fin coil or the structure is forced or induced draft.

Small diameter fans direct-connected to motors could be used on process exchangers but seldom are because too many of them are required and fewer larger diameter fans are equally efficient. When directed connected drives are used they are seldom field assembled due to their small shipping size. In air coolers mounted at grade, fans and motors are often mounted separate from the coil and structure to avoid transmitting vibrations between them. In process work, particularly in refinery and chemical plants, the fan and blower, positioned well above grade, are almost always structure mounted and in rare instances, are direct connected.

Fan drives in plate fin service are seldom field assembled because they are usually small in size; in process service fans larger than 14 ft in diameter are often field assembled. Preassembly and run-ins are preferred because this 20-min or so test assures clearances are correct. It is essential that fan and coil positions be maintained. V-belts stretch with use and are costly to replace. Some form of belt tension tightener is required.

A two-speed fan, when used, is generally furnished on only one drive per bay. Its use provides better airflow control, particularly when the ambient temperature is less than the design temperature.

In A/C duct systems, because of the large pressure losses that must be overcome, the blower system is assembled and purchased as a package that includes blower, motor and V-belt drive.

A two-speed fan, when used, is generally furnished on one drive per bay. This provides better airflow control, particularly when the ambient is less than the design temperature.

Chapter 7

Exchangers with Longitudinal-Fin Tubes

Longitudinal-fin tubes are in common use on two types of applications: heating or cooling of viscous liquids, and heating or cooling of low-pressure gases. They are also used in a variety of other applications. An advantage is that shell-side flow areas of longitudinal finned-tube units are large compared with typical shell-and-bare-tube units. Another is that the crossflow component is minimal using longitudinal fins except for those designs with punched holes in the fins. Both types are used in tank heater services.

7.1 DOUBLE-PIPE EXCHANGERS

The original longitudinal-fin unit is the double-pipe exchanger (Figure 7-1). Standard units are available using bare or finned tubes, and these are built to the rules of the ASME Code. Bare surface is usually built to pipe dimensions. Finned surface is normally attached to pipes but tubes could be used. Bare tubes are beneficial when both streams have roughly the same film coefficient. If the tube material needed is costlier than steel, it can be thinner if the tubes are not finned. In this case the bare tube surface will be built using the thinner tubes and not the thicker pipes. Finned surface is generally preferred when the heat transfer rate of one stream is 3 or more times that of the other. The fluid having the poorer heat transfer rate is likely to be on the fin side.

Double-pipe exchangers can be arranged in series or parallel, or a combination. Tube connectors have been standardized to accommodate various flow arrangements. An LMTD correction factor may be necessary but will usually be above 0.96. This is as efficient as most competitive designs. Standard units have shell design pressures of 650 psi or higher depending on temperature. Standard tube-side pressures are up to 1500 psi. Nonstandard units can be furnished for pressures well above these values (to 15,000 psi, for example). Double pipes are ideal where one stream has a large temperature drop that can be distributed over several sections in series. The use of interrupted fins can increase turbulence while at the same time improve the heat transfer rate (Section 14.4). Because this design is versatile and inexpensive, finned tubes or complete sections are ideal as spares. Units with the same size shells are interchangeable in terms of stacking units. As a rule, four standard steel units with $1^1/_2$-in-diameter steel pipes are less expensive than one code-constructed steel shell-and-tube unit. As sections

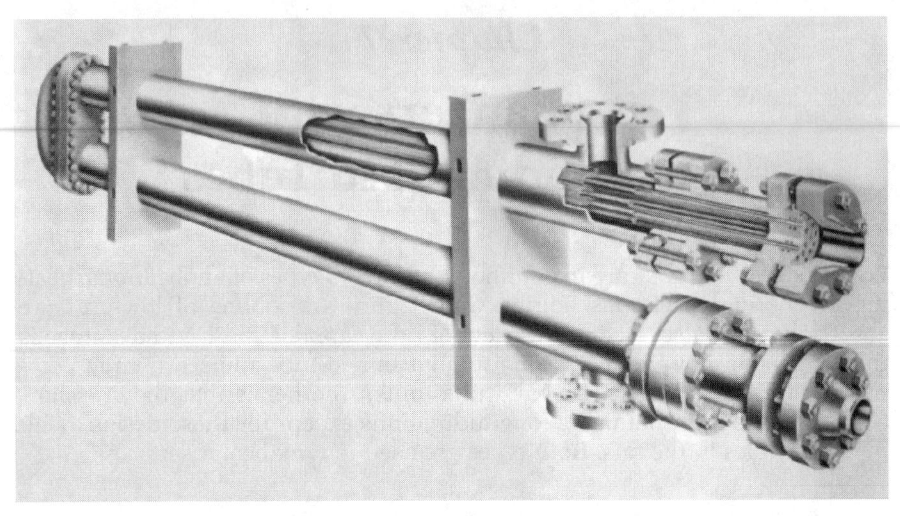

Figure 7-1. Cutaway diagram of double-pipe exchanger. (Reprinted by permission of Brown Fintube Company.)

are added, the cost advantage often switches to shell-and-tube units. This conclusion does not apply if large temperature crossovers exist or if the operating pressures on the shell or tube sides of a shell-and-tube exchanger exceed roughly 300 psi.

Double-pipe exchangers are available in many materials. Because they are standardized, steel units are often stocked or readily assembled, with quick shipment provided. This is a plus even when units are not stocked. Other code-stamped exchangers are usually not as readily available. In some designs fins are welded to tubes. In others, the fins are embedded in grooves, allowing the use of materials that cannot be welded to steel. If the fin material on steel tubes is changed from steel to copper or aluminum, the overall rate can be improved over that of steel fins and can result in the use of fewer sections. With liquids flowing, the percentage gain is usually in the 25 to 30% range, which can be used as a measure of the expected increase in heat transfer. Changing fin material is a good way of increasing the cooling (also heating) capacity of a unit. Be aware that this change may be limited because of the lower tensile stress of nonferrous materials. A calculation of fin tip temperature will probably be needed.

Double-pipe exchangers can be mechanically cleaned on either stream, and tubes are easily replaced. As flows increase, the shell diameter must be increased, but this requires higher fins to keep the tubes centered in the shell. If a plant expansion is under consideration, do not overlook adding sections in series or parallel with existing sections. If the flow is large or if the number of streams in parallel is excessive, consider using higher fins and larger-diameter shells to simplify manifolding and reduce pressure loss.

7.2 TANK HEATERS

Fuel oils and asphalts are difficult to pump when cold because of their high viscosity. To decrease this, products are heated. Two kinds of heaters are commonly used. One is a series of bare or spiral-fin pipes positioned near the base of a tank and heated by steam or hot water to keep the product warm (Figure 7-2). Pipes or coils are almost always carbon steel. Consider the possibility of a leak occurring. To find and plug it requires draining the system. Spare tanks are rarely available. To minimize the chance that a leak will occur, tank pipe coils are constructed using Sch. 80 pipe. Using this thicker material pipe provides a factor of safety against leaks as these pipes cannot be repaired without draining the tank. The shell and tube units described in the following paragraph do not need tubes this thick because, should a leak occur, the bundle can be removed, repaired, and replaced without draining the tank provided a shut-off valve is installed at the shell entrance. This design looks like the shell and tube unit shown in Figure 7.4 only installed in a tank similar to that shown in Figure 7-3.

Some shell-and-tube exchangers have longitudinal-fin tubes of steel. The bundle extends into the tank (Figure 7-3). In winter the tank temperature (not the exchanger's) is coldest at its periphery. At this boundary the fluid is near ambient temperature. When the fluid near the tank's periphery cools, it becomes more viscous. In doing so it will likely form a thick insulating layer around the tank that substantially reduces the heat loss from the liquid pool at the center of the tank.

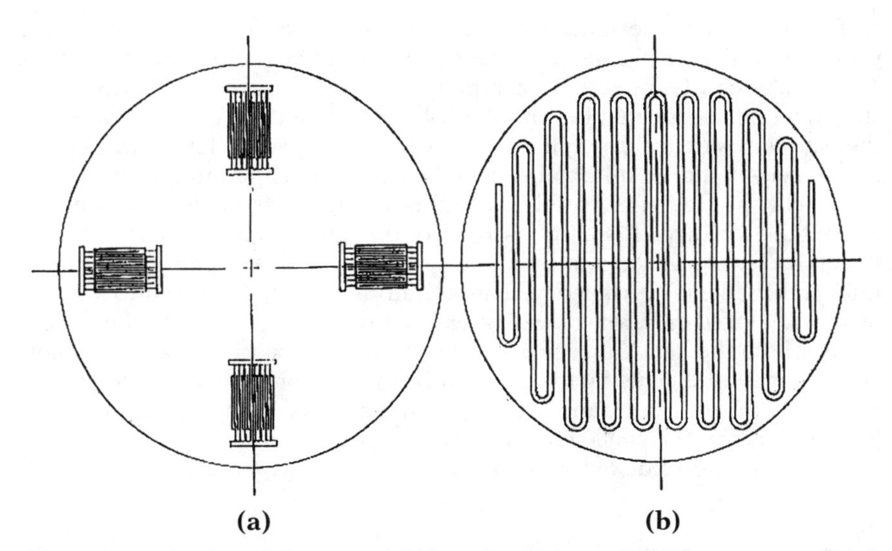

Figure 7-2. Tank coil layouts: (*a*) finned-coil layout; (*b*) layout using bare pipe. The advantages of finned coils are a need for fewer welds and less floor space, greater ease of cleaning, and the ability to isolate coils. (Reprinted by permission of Brown Fintube Company.)

Figure 7-3. Tank suction heater. (Reprinted by permission of Brown Fintube Company.)

Tank heaters are designed to project into this warm pool of oil. Their purpose is to warm the product sufficiently that it can be pumped from the tank. The bundle can be removed for repair if a shutoff valve is provided at the shell inlet nozzle located beyond the U-bends. Note that the oil stream is on the suction side of the pump. Hence, the pressure loss through the exchanger should be minimal to avoid cavitation at the pump. In selecting this exchanger, be sure to take advantage of the 2 ft of head (minimum) above the exchanger. Also, the viscosity of the fluid compared with its viscosity at the tube wall should not be ignored, as this factor substantially improves the heat transfer rate. This condition can be compared to a square of butter placed in a warm frying pan. At the hot surface, the fluid flows freely where the heat transfer is greatest though most of the butter remains viscous and solid. This factor can reduce the size of exchanger required by about two-thirds, a substantial saving in size and cost. Further, it reduces the pressure loss on the shell side, minimizing the chance of a vacuum developing on the suction side of the pump.

7.3 LINE HEATERS

The line heater (Figure 7-4) is an ideal application for a longitudinal-fin exchanger. It can be located on the suction or the discharge side of the

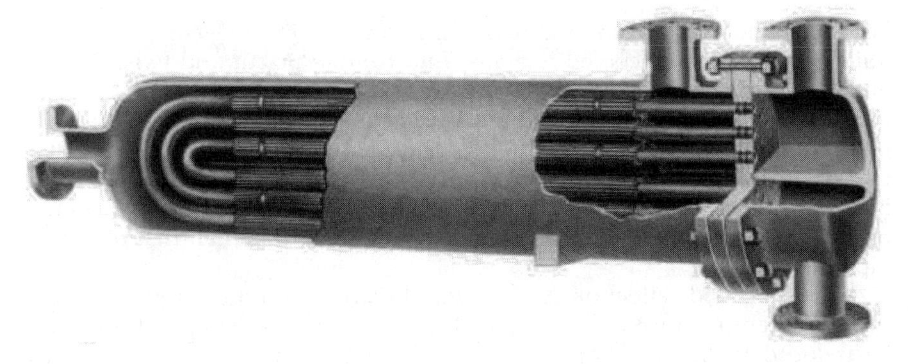

Figure 7-4. Line heater with longitudinal fins. (Reprinted by permission of Brown Fintube Company.)

pump. More pressure drop is available for its operation because its entering viscosity is less than that of the tank heater. The usual purpose of line heaters is to warm low-cost fluids such as asphalts, Bunker "Cs," and No. 6 fuels from about 75°F to 250°F. At this temperature these fluids can be vaporized at burner nozzles. Warming fluids to this temperature allows for the use of these lower-cost fuels.

7.4 OVERALL HEAT TRANSFER RATE USING LONGITUDINAL-FIN TUBES

Assume the fluid on the fin side of a longitudinal-fin double pipe exchanger is viscous and heated with steam. The viscosity at its tube wall surface will be different from its bulk viscosity depending on the temperatures at these points. Often the viscosity correction factor $(\mu/\mu_{\omega})^{0.14}$ is enough to reduce the size of exchanger required by half or more. Further, if the fins have cuts and twists or are designed with interruptions, the rate can be improved as these factors are additive. Another way of improving the rate is to change the fin to a non-ferrous material. The welded fin to tube design cannot be used to take advantage of this feature because these fin materials cannot be welded to steel.

The same exchanger, when used as a cooler, does not deliver the same results. The viscosity at the wall of the liquid being cooled will be greater than the bulk viscosity because its temperature is colder thus increasing the barrier to heat transfer at the surface. This can increase the size of cooler needed by two or more times. As in the heater case, interruptors or cut and twist fins can be used to improve the overall rate. These factors may be offsetting or one may influence the size of exchanger more than the other.

The same conditions apply to shell-and-tube longitudinal-fin exchangers in terms of heater and cooler wall viscosities. Overall rates are improved in heating services and decreased in cooling services. The reverse is true if the fluid being heated or cooled is a gas. Interruptors or cut and twist fins are not as effective in longitudinal fin shell-and-tube exchangers because too much shell side flow area per tube exists that allows the viscous fluid to bypass surface. Support baffles holes are drilled to a slightly larger OD than that of the finned tubes. The viscous fluid will pass the baffles at these openings thus improving the overall rate. In shell-and-tube longitudinally-finned units the length of shell-side flow will usually be less than half that of double pipe exchangers.

A beneficial side effect of using longitudinally-finned tubes follows. When the metal temperature is too high hydrocarbons can, by temperature alone, be reduced to lighter components which is rarely beneficial and often a handicap. One way of reducing metal temperature is to use finned rather than bare tubes. The metal temperature of heaters with steel tubes and fins contacting viscous hydrocarbon fluids will usually be about 10 to 15°F less than when bare tubes are used. In most cases the use of finned tubes avoids the problem.

Chapter 8

Plate Exchangers

Plate exchangers are constructed of nearly rectangular plates rounded at the corners and separated by gaskets located near their edges (Figure 8-1). Their assembly is shown in Figure 8-2. The volume bounded by two or more plates and their gaskets contains the flowing fluids. This applies to each side of each plate. The area within these boundaries is heat transfer surface. This construction is ideal in many applications. Several improvements have been made in the basic concept to enhance the heat transfer characteristics of this construction. Seal welds are sometimes used, omitting the need for gaskets. Other improvements will be described in this chapter.

8.1 CONSTRUCTION FEATURES

Plates are manufactured by stamping. Production techniques have improved so that more metals are available as plate materials. Plate metals include stainless steel, titanium, nickel alloys, and specialty metals such as tantalum, plus others.

During manufacture grooves are stamped in plates to contain gaskets. The surface is often modified in a herringbone fashion (see Figure 8-3) stamped to aid in creating turbulence, and to break up streams in laminar flow. The stamping depth (visible in Figure 8-4) is selected based on the quantity of fluid flowing, the allowable pressure drop, and, particularly, the size of particles, such as pulps, being carried by the fluid. Plates can be stamped to larger depths, but particle sizes must be known before plates can be guaranteed to work. In the installed position, plates touch each other at many places, a feature that adds stability when pressure is applied to either stream.

Crimping increases a plate's ability to sustain pressure. Several plate thicknesses are available, which vary with the material used. The number of gasket and glue materials available for seals has increased. In some designs glues are not needed to hold gaskets in position. Other improvements have been made that increase the number of applications where plate exchangers can be used. The current pressure and temperature limits are 400 psi and 375°F. The pressure and temperature limits are not usually reached simultaneously. Often the limit is a function of gasket material.

8.2 PLATE EXCHANGERS VERSUS SHELL-AND-TUBE AND DOUBLE-PIPE EXCHANGERS

Plate exchangers are preferred over shell-and-tube or double-pipe units in some applications. They are well suited to counterflow construction, thus

Figure 8-1. Fully assembled plate exchanger. (Reprinted by permission of Alfa Laval Thermal, Inc.)

maximizing the LMTD correction factor. Their construction minimizes the need for welding. There are other pluses as well. Consider a 700 ft^2 shell-and-tube exchanger with 474 tubes $^5/_8$ inch in diameter and 9 ft 0 in long (refer to Table 2-1). If this surface were in a 21-in-diameter shell-and-tube exchanger, its volume would be 21.64 ft^3. A plate exchanger with fifty-five 71-

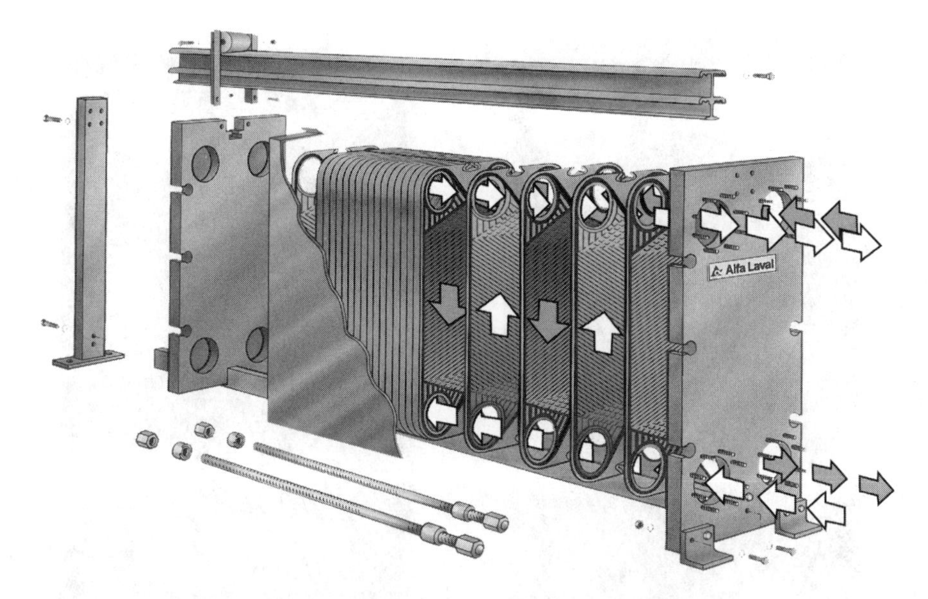

Figure 8-2. View of plate exchanger showing component parts and internal flow. (Reprinted by permission of Alfa Laval Thermal, Inc.)

by 33-in plates also has 700 ft^2 of surface but occupies only 8.2 ft^3. Because of their reduced volume, plate exchangers are competitive with shell-and-tube units even though built of costlier materials. Check the thickness of material required before confirming this conclusion, as it does not apply in all cases. Cylindrical tubes are thicker and can sustain much higher pressures and temperatures.

Manufacturers of plate exchangers speak of turbulence created by the herringbone-shaped plates as significantly improving the heat transfer rate. The question arises, How can the heat transfer rate in the turbulent zone be improved without increasing pressure loss? As there is no quarrel with the conclusion, the following is offered as an explanation of why the heat transfer rate is improved using this construction. Often the goal is to continuously break up laminar flow, which the herringbone design accomplishes.

Shell-and-tube and double-pipe exchangers often use tubes 20 ft long. In service they are frequently arranged in 2- or 4-pass units, meaning a flow length of 40 or 80 ft. By contrast, a plate exchanger's flow length per pass is seldom longer than 8 ft and is usually less. Further, the numbers of passes in plate exchangers are usually more than in a corresponding shell-and-tube unit. The reason for this is best explained by referring to Table 2-1. The equivalent diameter of plate exchangers is usually less than the diameter of any tube used in Table 2-1. This means that plate lengths are relatively short. If a shell-and-tube selection has an allowable pressure drop of 10 psi and a straight-travel loss of 1.25 psi, the pressure drop limit will not be met if the

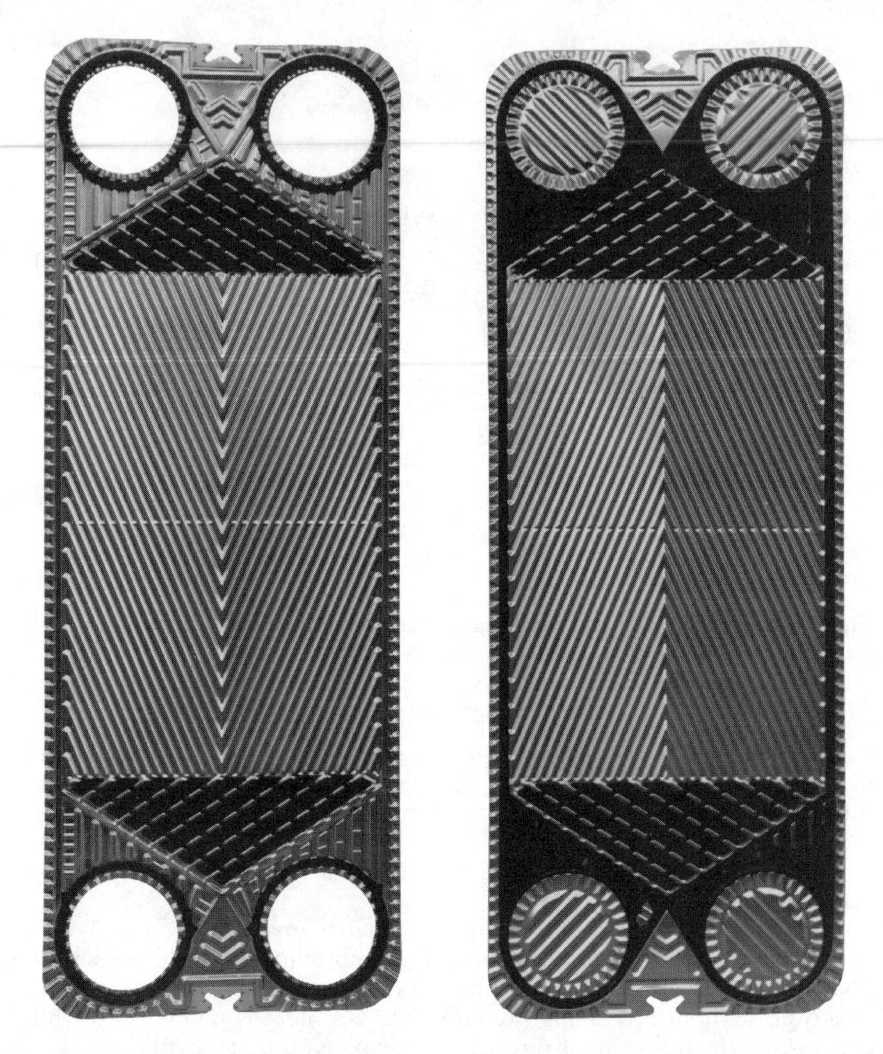

Figure 8-3. Typical design of a plate exchanger surface. (Reprinted by permission of Alfa Laval Thermal, Inc.)

number of passes is increased from 1 to 2 or from 2 to 4, because this change increases the pressure drop by 8 times, not including turning losses. The length of flow in plate exchangers is less than for other designs, resulting in a reduced pressure drop per pass and thus allowing more passes to be used or a higher pressure loss per foot of travel. A 4- or 6-pass exchanger might have the same flow length as a 2-pass shell-and-tube unit. Increasing the number of passes from 4 to 6 or from 6 to 8 increases the pressure drop, respectively, by $(6/4)^3$, or 3.38 times, or by $(8/6)^3$, or 2.37 times, a substantial reduction

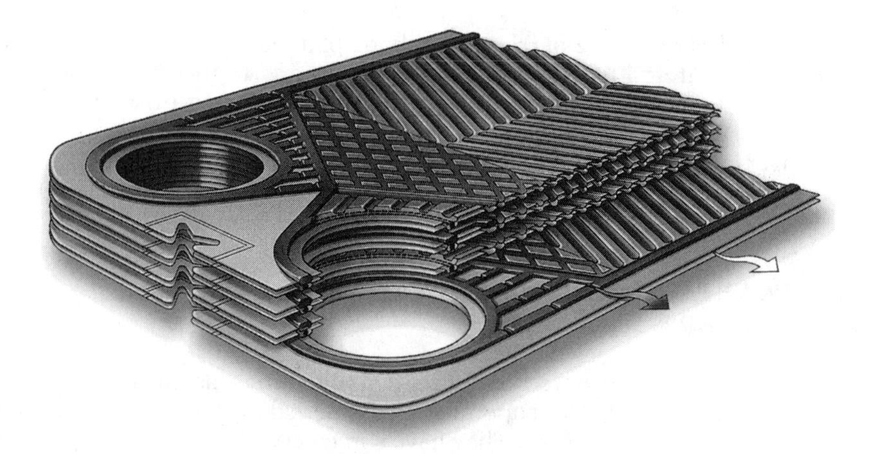

Figure 8-4. Double-wall heat exchanger plate. (Reprinted by permission of Alfa Laval Thermal, Inc.)

from an eightfold increase. The result is more flexibility for improving the heat transfer rate and better use of pressure drop. A lower pressure drop can be obtained by reducing the number of passes from 8 to 6. This geometry is one reason why plate exchangers provide more pass choices and higher rates on both streams than do other exchanger designs. Another reason is given in Section 10.2.2. A third is that plate thicknesses are thinner resulting in a decrease in metal resistance. A fourth is that the herringbone shape breaks up any laminar or transition flow component present thus acting like a turbulator. The conditions are additive in improving the heat transfer rate. Moreover, these pluses can and often do occur on both streams.

Small shell-and-tube exchangers in light industrial service have much higher overall heat transfer rates than large shell-and-tube exchangers. This variation can be explained on the basis of their having a higher pressure drop available per foot of travel. In plate exchangers this advantage is greater because it applies to both streams. Thus, heat transfer rates are even further improved.

The tube-side pressure drop in shell-and tube exchangers can be varied by changing the tube diameter, tube length, or number of passes. In plate-fin units, the hot and cold fluid heat transfer rates and pressure drops can be varied by changing the length and width of plates, the number of passes and plates, and the depth and shape of herringbone stampings. Each is controlled by the heat transfer engineer. In total, these features provide the engineer with more variables to maximize the selection.

Both streams of plate exchangers have their flows interrupted at short intervals of travel by herringbone shaped plates. This is similar to the feature that makes turbulators and cut and twist fins efficient at improving the heat transfer rates in shell-and-tube, double-pipe, and some air coolers. However, there is a difference. The herringbone shape plate is used in nearly all plate exchangers. Cut and twist fins and interruptors are, in the main, used only on streams that are in laminar flow or that are operating in the transition region. It is under these conditions that low L/d ratios are efficient in most exchangers.

Fouling in various types of heat exchangers is discussed in Chapter 10.

8.3 ADVANTAGES AND DISADVANTAGES OF PLATE EXCHANGERS

Plate exchangers are preferred in such applications as those breweries, dairies, distilleries, HVAC, papermaking, winemaking, and water services and in marine and other uses. They have some very distinct advantages. For many services, more materials of construction are available, which can be an advantage over shell-and-tube or double-pipe units. The plate exchanger can have dual plates to separate the streams and avoid any mixing between them (Figure 8-4).

Plate exchangers approach counterflow conditions in most applications. Temperature crossovers of 10 to 100°F are common. It is seldom necessary to use several plate exchangers in series rather than one large unit to account for an LMTD correction.

Plate exchangers often have a competitive advantage when the available pressure drop is high (5 to 10 psi) and when most of it is used in selecting the exchanger. The reason is the plate exchanger ís relatively short length of travel (due to its low equivalent diameter, Table 2.1) resulting in more tube side pass choices (Section 8.2). The result is higher heat transfer rates on both streams. Consider the selection in terms of competitor's offerings. The plate exchanger is smaller in volume and has less surface than competitive shell-and-tube or double-pipe units. This can be a cost advantage as surface could be added during a plant expansion by increasing the number of plates without changing heads or seals. This may not be true for a given application but can be true provided the unit is selected initially for both design and alternate conditions (see Section 8.4, Planning Ahead). Its pluses are that it occupies less space, is made of more expensive materials, and is easily cleaned on both streams. Before finalizing a selection it is a good idea to check if, when passing through the exchanger, the flow changed from laminar to transition or turbulent. If it did, the exchanger should be rerated, as necessary, to satisfy operating needs. See also Section 8.2.

On the other hand, there are also several disadvantages to the use of plate exchangers. Plate exchangers are not available with extended surface other than that of stretching that occurs in the stamping process. They are not recommended when fluid particles are larger in diameter than the stamped depth of the plates (see Figure 8-4). Larger particles will plug the flow path. When particles are present, spiral exchangers (not covered in this book) are recommended, or plates stamped to a greater depth.

The pressure and temperature limits of plate exchangers are low when compared with shell-and-tube or double-pipe units, roughly 400 psi and 375°F. Competitive designs can be built to 15,000 psi and 1000°F or more. These limits are not needed often and are seldom reached simultaneously.

Plate exchangers are not readily adaptable to low-pressure vapor or gas cooling services, although it should not be assumed that they are never used in these services. In fact, a standard line of plate exchangers has been developed for this purpose. Most refrigerants operate above atmospheric pressure where these units are suitable to ideal. Their seals are not gaskets but welds with their principal benefits being higher quality metallurgy and less space required than for competitive units. Shell-and-tube exchangers are commonly used in gas, vapor, and liquid services, including those with partial or complete condensing while plate exchangers and double-pipe units are more common in liquid services only.

The magnitude of fouling has a strong effect on the size and cost of plate exchangers. In general, the plate exchanger is less efficient in high-fouling services, though plate exchangers can be easily cleaned.

In industrial plants, avoiding fire is an ever-present concern. Because plate exchangers have a low design temperature, about 375°F, an unwanted fire in the plant may raise the external temperature sufficiently to allow the product in an exchanger to feed the fire. Because of their low temperature limit, plate exchangers are not as common as other designs in refineries and chemical plants. This limitation usually does not apply in nonflammable applications.

Instrumentation is often installed in piping near a plate exchanger's inlet and outlet. In shell-and-tube exchangers, these connections are installed in the exchanger's nozzles and not the mating piping provided they are a requirement. There are problems associated with installing some exchangers or maneuvering them into position. Plate exchangers can be disassembled, taken through doorways piece by piece, and reassembled without difficulty. While they cannot be operated at pressures as high as other exchangers, this condition is generally not a factor in jacket water and lube oil cooling and in many other services. The low-pressure side of plate exchangers should be turned on first. Be aware that pressure surges and thermal shocks should be avoided, as plate exchangers are not ideal for these conditions. In most applications this is not a problem.

8.4 PLANNING AHEAD

At the time of the initial selection, exchangers should be sized for present and future conditions. Even when a new exchanger is part of a plant expansion, there is no assurance that it is the best choice unless both present and future conditions are considered initially. Inlet and outlet nozzles should be sized for future conditions, if known, at the time of selection. Under favorable conditions plate exchangers can accommodate a plant expansion merely by adding plates with no other changes needed.

If particles are present they may block the flow path. This problem can be solved by using plates with a greater stamping depth. The particle sizes should be known before the unit is selected.

8.5 COST OBSERVATION

Consider the inside surface of tubes in shell-and-tube exchangers. Many seals are needed to contain this surface, including those for tube-to-tubesheet connections and for connections at the floating head cover, floating head flange, floating head backing device, stationary head channel, and channel cover and channel flanges. Connections cost money. Now consider seals in plate exchangers. Flow is contained by heavy end plates, tightening rods, and gaskets. If plate exchangers are viewed in terms of price versus shell-and-tube exchangers, it should be apparent that plate exchanger seals will be less expensive than tube-side seals for shell-and-tube units. Compare the effort needed to construct the shell side of shell-and-tube exchangers with that of constructing plate exchangers. In one, the needed materials include tie rods, baffles, shell, shell cover, seal strips, shell flanges, and bolting. In the other, no additional material is needed other than gaskets. If special materials are needed, plate exchangers can use the same material for both streams, which can result in a considerable cost advantage. Shell-and-tube exchangers, by comparison, often have different materials in each stream.

8.6 A SEASONAL APPLICATION

Consider the streams of water cooled chillers. In one, water is cooled in the evaporator which is in a closed loop system. This water eventually flows to air-handling unit(s) where building heat is removed. In the other cooling stream water flow is from the cooling tower to the chiller's condenser. At no time do the water streams mix. Both chiller and cooling tower(s) are expensive items and, for long service, should be properly maintained.

The best way of accomplishing this is to have both streams analyzed by water treatment specialists to determine their physical and chemical properties. Their recommendations on water treatment should be followed if chiller and cooling tower efficiencies are to be maintained. Through control of purity the chiller and cooling tower will be subject to less corrosion, deposition of heat-resistant scale, organic growth, and sedimentation. Avoiding these problems is a way of meeting performance and reducing operating and maintenance costs. Normally both water streams are treated for this purpose.

The fact that water treatment can reduce fouling is no guarantee that it *will* reduce fouling. Some waters cannot be corrected by treatment alone. In such cases additional fouling protection should be provided when sizing evaporators and condensers. In most applications, treating water improves the operation of a system.

The question should be asked, What this has to do with heat exchangers? Consider the following. Cold water leaves the tower at about 75°F in the summer and as low as 33 or 34°F in the winter. With large quantities of cold water available during the colder months, running the chiller may not be necessary. Energy saved by turning it off is often enough to pay for a heat exchanger that serves the purpose of the chiller but operates in place of the chiller in the colder months.

The exchanger (without the chiller) is used during those periods when the tower water temperature is below 50°F. In the exchanger the closed loop chilled water flows in one stream and cold water from the cooling tower in the other. The cooling tower water provides the same or more cooling than does the chiller but at a lower cost because the chiller will be idle. In this service the plate exchanger offers many advantages. Water conditions in both streams are controlled, although there is no guarantee that they always will be. Heavy rains, low water levels in lakes and rivers, or local digging and construction can change the contaminants entering a system. The exchanger can be cleaned on both sides if necessary. Keeping the chiller clean by using treated water is a benefit for the plate exchanger, because it can be sized using a reduced fouling factor.

Both streams use materials other than carbon steel. Plate exchangers are likely to be installed in the mechanical room, where space is at a premium, making their small volume advantageous. Another benefit is that required cleaning space is less than for other exchangers. In this application the log mean temperature difference is low and flows are high. Plate exchangers are ideal for both conditions. These exchangers do not run continuously, that is, 8760 hours per year. Their cleaning schedule can be arranged without the need of meeting a deadline, to coincide with times when the units are off-line. Both streams have nearly the same heat transfer coefficient. This is a plus because plate exchangers do not feature extended surface. A water leak to ground should not be a problem because no fire hazard exists.

Chapter 9

Comparing Exchanger Types

The preferred construction for an application might be a shell-and-tube unit, a double-pipe design, or a plate exchanger. The most economical unit is apt to be determined by the design pressure or the design temperature. They differ for each design. An estimate, based on experience, is that 80% of exchangers operate at pressures under 300 psi. Sometimes a quick estimate is needed to decide which exchanger to select. The following tabulation gives a rough comparison of standard units:

Shell-and-tube exchanger	75 psi at 375°F (packed head construction)
	150 psi at 650°F
	300 psi at 650°F
Double-pipe exchanger	650 psi at 650°F (shell side)
	1000 psi or higher and 1000°F or higher, but usually not simultaneously (tube side)
Plate exchanger	400 psi at 375°F (both streams)
Air coolers	(Discussed in Section 9.1)

Shell-and-tube exchangers can be operated at higher pressures and temperatures on either stream than those listed, but these would usually be nonstandard units, significantly higher in cost. Operating pressures or temperatures of double-pipe exchangers, on the other hand, can be raised from those shown for only a modest increase in price. The cost of manifolds should probably be added to double-pipe costs (when comparing units on economical grounds), as more connections are normally needed than in competitive designs. The point is that one exchanger may be a better economic choice than another based on design pressure and temperature, space considerations, metallurgy, cleaning, delivery time, manifolds, or other considerations.

9.1 AIR COOLER HEADER LIMITS

The design temperature and pressure leeway that exists for shell-and-tube and double-pipe units does not necessarily hold for air coolers. Operating pressure is more likely to dictate the construction to use. Standard box headers with shoulder plugs can generally be used to 1000 psi and 600°F. Headers can be modified to contain substantially higher pressures. Coverplate header pressures are limited by the number of tube layers (see Table 4-2). These can vary with the use of smaller or larger tubes, but the headers' operating pressure limit is much the same as shown in Table 4-2. Units with welded U-bends (see Figure 2-4) can perform at higher pressures and temperatures than shoulder-plug or cover-plate headers, to 15,000 psi if necessary. These are almost always one-of-a-kind items. Brazed U-bend units may be operated at higher pressures than cover-plate units even though these are built of softer materials and thinner tubes.

Plate-fin exchangers have lower design pressure and temperature limits than spiral-fin units, roughly 250 psi at 200°F. Generally, they do not have a floating inlet header, as it is not needed in most applications (spiral-fin units can have a floating inlet header but generally do not). They are often built with materials of low tensile stress, that is, nonferrous tubes. This construction is ideal in low temperature and pressure applications (lube oil and jacket water coolers, mainly), where they are frequently used. The market for this product is higher in dollar value than that for process exchangers. These applications are repetitive and do not require the broad skills needed for selecting process exchangers. It does not follow that their selection is less important.

9.2 LIMITATIONS IN THE SELECTION OF FINNED SURFACE

Most extended-surface constructions were developed to satisfy the needs of specific applications. They were not built with the goal of determining where they could be used. Manufacturers always seek new applications for their products. Many surface types are available, but all have limitations. The goal is to select the best surface for the application. Here are reasons for choosing one design over another.

One of the best applications for longitudinal fins is for liquid flow in paths parallel to fins. Similarly, low-pressure gases and vapors flow across spiral- or plate-fin surface in paths perpendicular to tubes to take advantage of the short length of travel and large flow areas. While often true, these generalizations do not always apply. Spiral lowfins are beneficial in both long and crossflow. Viscous fuels in tanks are heated in crossflow using spiral-fin coils. Low-pressure gases and vapors flow parallel to longitudinal fins in shell-and-tube units because of the low pressure drop that results. Crossflow offers a large flow area and short length of travel, a plus when the pressure drop must be small. Longitudinal fins are beneficial when the operating pressure must be contained, an advantage when they are used in shell-and-tube units. Liquid flow needs are the opposite. Liquid flows are dense, requiring less flow area, and are nearly always under pressure, so that more pressure drop is usually available, resulting in improved heat transfer rates.

When seeking the benefits of crossflow and the pressure-containing advantages of shell-and-tube units, do not overlook the model J design when air or gases are present (refer to Figure 3-2). This design offers numerous advantages:

1. Large dome areas can be furnished, minimizing the inlet nozzle pressure loss.
2. Large dome areas provide good distribution across a coil.
3. The length of travel across a bundle is minimal.

4. The area for gas flow is maximum, thus minimizing crossflow pressure drop.
5. Inlet and outlet nozzles can be positioned to reduce piping needs.
6. In compressors, a low first-stage pressure loss means a higher-pressure output at the third and/or fourth stages.
7. The surface can be plate-fin, spiral-fin, lowfin, or bare-tube.

Between the extremes of operating pressure, density, flow, and pressure drop, either cross- or long flow may be the best choice. The point is to have an open mind when selecting surface.

Chapter 10

Fouling Factor Considerations

After exchangers have been in service, dirt and scale collect on their surfaces, usually at different rates, thus reducing the amount of heat that can be transferred, or in other words, reducing the exchanger's performance. As a practical matter it is not reasonable to clean an exchanger every time some scale buildup occurs. Rather, it is customary to minimize the problem by adding surface and attempting to anticipate the amount of buildup that will occur on both streams over a period of time while still maintaining performance. This is done by adding another resistance to each stream when calculating the overall heat transfer rate. These resistances are called fouling resistances or fouling factors. Their suggested values have been established using industry experience and operating conditions as a guide. The question remains, How much surface should be included? Some answers will be provided in this chapter via a look at plant conditions followed by examples.

10.1 PLANT CONDITIONS

In the first instance, a heat exchanger is installed in a refinery. If only minimum fouling is accounted for in its sizing, the unit will foul in a short time and be incapable of transferring the needed heat. One way to correct this is to shut down the exchanger (translation: the *refinery*) so the unit can be cleaned. This doesn't make sense, because it results in a loss of production. A more reasonable approach is to design sufficient fouling into the unit so that it will run one to two years or more without cleaning. Refineries, chemical plants, and power plants follow this procedure. Exchangers can be cleaned following shutdown. For sizing purposes the recommended fouling factors for continuously operated units provided in the Standards of the Tubular Exchanger Manufacturers Association (TEMA) are given in Table 10-1.

A similar situation is faced in large steam power plants, that is, determining how much fouling should be allowed for in the design. These plants are designed to be onstream for longer periods of time than refineries. They may run five years or more without a shutdown. The quantity of tubing required is high, and it is mainly of nonferrous materials. Tubes are costly compared with steel, the metal commonly used in refineries. Power plant water is almost always recirculated, cleaned, and treated. This exchanger's fouling rate can also be reduced by cleaning the tubes while the plant is operating (Section 14.5). For these reasons fouling resistances are usually less in power plants than in refineries. With sufficient surface supplied to accommodate fouling it is not necessary to shut down for cleaning except at extended time intervals.

Table 10-1 Recommended Fouling Resistance for Water

If the heating medium temperature is over 400°F and the cooling medium is known to scale, these ratings should be modified accordingly. All units are hr °F sq ft / BTU.

	Water velocity, ft/s			
Temperature of heating medium	**Up to 240°F**		**240 to 400°F**	
Temperature of water	**125°F**		**Over 125°F**	
	3 and less	**Over 3**	**3 and less**	**Over 3**
Seawater	0.0005	0.0005	0.001	0.001
Brackish water	0.002	0.001	0.003	0.002
Cooling tower and artificial spray pond:				
Treated makeup	0.001	0.001	0.002	0.002
Untreated	0.003	0.003	0.005	0.004
City or well water	0.001	0.001	0.002	0.002
River water:				
Minimum	0.002	0.001	0.003	0.002
Average	0.003	0.002	0.004	0.003
Muddy or silty	0.003	0.002	0.004	0.003
Hard (over 15 grains/gal)	0.003	0.003	0.005	0.005
Engine jacket	0.001	0.001	0.001	0.001
Distilled or closed cycle				
Condensate	0.0005	0.0005	0.0005	0.0005
Treated boiler feedwater	0.001	0.0005	0.001	0.001
Boiler blowdown	0.002	0.002	0.002	0.002

Source: Reprinted by permission of the Tubular Exchanger Manufacturers Association (TEMA).

A/C systems and commercial exchangers do not have as much fouling allowance built in as refineries, chemical plants, or power plants. The latter operate continuously, that is, 8760 hours per year. Most A/C systems and commercial applications operate about half this time or less. Hence, their potential for fouling is less. Fouling can build up when a unit is idle. In power plants and mechanical rooms, steps are taken to furnish clean water to chillers and other equipment. Clean water will reduce fouling on the chiller and cooling tower, and also on other equipment in its stream such as the evaporator, condenser, and air handling unit. A/C systems are often idle in the evening, at night, or on week-ends. Cleaning can be done at these times. Under these conditions, fouling has a minimum effect on the availability of equipment.

The subject of fouling is addressed before equipment is selected or built. If it becomes a problem after equipment is installed, it is usually a trouble job. These fouling conditions should be considered before selecting equipment as all resulted in trouble jobs during the author's career.

Some streams contain particles. Users should know their size, number, and effect on the system. Sometimes these particles act like sandblasters cleaning tubes when passing through the exchanger. The result is a minimum fouling resistance need. Users should furnish particle information. Consider another stream with particles entering the tubes at high velocity. Under this condition, local pitting often occurs causing failures at tube entrances. Velocity limits at this location should be defined. Brush cleaners, when feasible, can effectively reduce fouling resistances and the size of exchanger required (Section 14.5). Do not overlook the possibility of large particles blocking flow effectively creating an undersized exchanger (Section 15.2). This is fouling of another sort. Should a fouling condition be anticipated, make sure the exchanger selected has design features that correct the problem.

Exchanger surface is usually measured on the outside of tubes though sometimes it is measured on the inside. Consider a 1-in 12 BWG tube. Its inside area is (0.205 sq ft / ft), its outside area is (0.262 sq ft / ft), and with fins added, the outside surface area might be, as an example, (1.540 sq ft / ft). In this case the external surface of the finned tube will be (1.540 / 0.262) or 7.51 times the surface area of the inside of the tubes. The heat that passes these surface areas is the same so the decision is where to measure exchanger surface. In this book surface is measured on the outside of tubes and includes the surface area of fins present, if any. When water or any other fluid is in the tubes its heat transfer resistance is usually not in direct contact with the surface used to size the exchanger. Hence, the resistance at this surface is modified by the (Ao/Ai) factor to determine its equivalent value where the exchanger's size is measured. The correction applies to tube side fouling for similar reasons. As an example a 1-in 12 BWG bare tube with an inside fouling resistance of .003 (hr °F sq ft / BTU) has a fouling resistance equal to 0.0038 (hr °F sq ft / BTU) when transferred to the outside surface. If fins are added to the outside surface, the (Ao/Ai) ratio in this case will be increased to (7.51)(0.003) or (0.0225 hr °F sq ft / BTU). This ratio of resistance correction relative to the total resistance does not affect the overall rate in the same way for all exchanger designs. There are times when a poor choice of fouling resistance is enough to make a selection non-competitive. Be aware that when fins are added the metal resistance also changes.

Tables have been developed to show the influence of fouling resistances on heat exchanger designs. The effect of fin side fouling will usually be less which will be illustrated by example.

10.2 FOULING EFFECT ON VARIOUS EXCHANGER DESIGNS

To account for fouling, surface is usually added to exchangers to assure they will perform at all operating conditions. Fouling resistances are defined by the engineer or established using industry standards or recommendations (Table 10-1). The percent increase in surface, assuming the same fouling resistances, is not the same for competitive designs as will be illustrated by examples. These conditions affect the surface needed for fouling.

1. Distribution, amount assigned to each stream.
2. (Ao/Ai) ratio, that is, area on the outside of tubes relative to that on the inside including equivalent diameter situations. These differ for plate, shell-and-tube and double pipe exchangers.
3. The amount of external or internal finned surface, if any.
4. The value of the fouling resistance relative to the heat transfer resistance for the same stream.
5. On which side of an exchanger should a stream be placed? It is easier to fin external than internal surface.

10.2.1 Plate Exchangers

Two examples of plate exchanger resistances, r, are given in Table 10-2. These differ in the way that fouling resistances are distributed. The table indicates the fouling resistances on both streams can be totaled (in the example, 0.0005 + 0.0015 or 0.001 + 0.001). The end result is the same because the (Ao/Ai) factor is 1.

It is not good practice to design an exchanger with a clean rate, U, that is more than doubled the fouled rate. In this case the the clean resistance, R, is 0.0011 (0.0004 + 0.0004 + 0.0003); its U is 909. The fouled rate is 322. Thus about 2/3 of the cooler's surface is furnished to account for fouling resistances. This could be considered as designing for fouling rather than for heat transfer and is considered poor practice. With these fouling resistances included in the design the unit could, when clean, transfer so much heat that it could overheat or overcool a product or leave set-point temperatures difficult to control. This situation does not occur frequently but the rater should be on the lookout for it. It is as common in shell-and-tube applications as in plate exchangers but it rare in double pipe exchanger applications.

Assume in this example that the shell side resistance was 0.004 instead of 0.0004, a not uncommon scenario. In this case the clean resistance, R, is 0.0047 (0.004 + .0004 + 0.0003); the corresponding 1/R or clean U is 213. The

Table 10-2 Plate Exchanger Rates Resulting from Different Fouling Distributions

Heat transfer rate U in units of Btu/(h · ft^2 · °F); R or r units are hr °F sq ft / BTU

Resistance *r*	**Case 1**	**Case 2**
Hot side	0.0004	0.0004
Cold side	0.0004	0.0004
Metal, stainless	0.0003	0.0003
Hot-side fouling	0.0005	0.0010
Cold-side fouling	0.0015	0.0010
Total resistance R	0.0031	0.0031
Heat transfer rate $U=1/R$	322.6	322.6

fouled R is 0.0067 (0.0047 + 0.002) and the corresponding clean U is 149. Hence, 43% excess surface [(213 – 149)/149] is provided and the surface required is not doubled by fouling resistances.

These examples illustrate that fouling resistances are sometimes influenced by their magnitude relative to other exchanger heat transfer resistances. To account for fouling the surface needed is not well defined by specifying a percentage increase in surface. In the examples the increases were 43% and 182% [(909 – 322)/322], yet the fouling factors in both examples are the same. A percentage increase in surface is usually specified for small commercial exchangers only.

When determining fouling resistances, consider the plant's fouling history and whether its operation is continuous or intermittent. If the latter, a lower design fouling resistance may be feasible because the unit can be cleaned when it is not operating. In these and in examples and tables following that apply to fouling, metal and fouling resistances are modified to their equivalent values at design surface areas. This has been done to show the total effect that fouling has on competitive exchanger offerings.

10.2.2 Shell-and Tube Exchangers

Three representative sets of resistances in a shell-and-tube exchanger are listed in Table 10-3. Total fouling resistance is the same in all cases except that it is distributed differently as outlined in the table footnotes. In shell-and-tube exchangers the fouling resistances cannot be totaled, as they can in plate exchangers, because a surface ratio correction factor is needed. For comparison purposes, using Table 10-3 as a reference, the total fouling resistance for each example has been modified to the outside surface of the exchanger with these results.

Table 10-3 Shell-and-Tube Exchanger Rates Resulting from Different Fouling Distributions

1-in (12 BWG) Tubes

Heat transfer rate U in units of Btu/(h · ft^2 · °F); R or r units are hr °F sq ft / BTU

Resistance r	**Case 1**	**Case 2**	**Case 3**
Shell side	0.00400	0.00400	0.00400
Tube side, outside	0.00120	0.00120	0.00120
Metal, steel	0.00041	0.00041	0.00041
Shell-side fouling	0.00200	0.00000	0.00300
Tube-side fouling (A_o/A_i included)	0.00128	0.00380	0.00000
Total resistance R	0.00889	0.00941	0.00861
Heat transfer rate $U=1/R$	112.5	106.27	116.14

Notes:

Column 1: Fouling 0.002 on shell side, 0.001 on tube side.
Column 2: Fouling 0.000 on shell side, 0.003 on tube side.
Column 3: Fouling 0.003 on shell side, 0.000 on tube side.

Case 1 (0.002 + 0.00128) = 0.00328
Case 2 (0.000 + 0.0038) = 0.0038
Case 3 (0.003 + 0.0000) = 0.003

In all cases the total uncorrected fouling is 0.003. However, the sum of fouling, heat transfer, and metal resistances must be totaled to determine the overall resistance and rate. Some interesting conclusions derive from the table. When all fouling is on the tube side, the overall heat transfer rate in this example is 9% less than when all is on the shell side. This means the exchanger must be 9% larger. Another conclusion is deduced by comparing this table with Table 10.2 relating to plate exchangers. Since the (Ao/Ai) factor for plate exchangers is 1, the tube side fouling and heat transfer resistances are not increased in value as they will be using the cylindrical tubes of shell-and-tube and double pipe exchangers. Hence, this is one reason why higher overall heat transfer rates for the same application can be attained using plate exchangers than for any comparable bare tube exchanger design.

10.2.3 Double-Pipe Exchanger

When the tube side fluid has a high heat transfer rate or low heat transfer resistance compared to that of a higher heat transfer resistance stream on the outside of the tube, fins are known to be beneficial. When the fouling resistance on the fin side is high compared to the fouling resistance on the tube side, a similar benefit accrues. Hence, fins can be beneficial for both conditions and these are additive.

The example of Table 10-4 is similar to that of Table 10-3 except that in all cases the total fouling is 0.006. Also in all cases, the Ao/Ai ratio is 7.5:1. Note how the heat transfer rate of the exchanger changes due to the distribution of fouling. As might be deduced from the table, high fouling resistance on

Table 10-4 Double-Pipe Exchanger Rates Resulting from Different Fouling Distributions, h · ft · °F/Btu

Ratio of A_o to A_i = 7.5:1. Heat transfer rate U in units of Btu/(h · ft^2 · °F).

	Case 1		Case 2		Case 3	
Resistance r	**Clean**	**Fouled**	**Clean**	**Fouled**	**Clean**	**Fouled**
Shell side	0.0300	0.0300	0.0300	0.0300	0.0300	0.0300
Tube side × A_o/A_i	0.0075	0.0075	0.0075	0.0075	0.0075	0.0075
Metal, steel	0.0220	0.0220	0.0220	0.0220	0.0220	0.0220
Shell-side fouling		0.0050		0.0060		0.0000
Tube-side fouling × A_o/A_i		0.0075		0.0000		0.0450
Total resistance R	0.0595	0.0720	0.0595	0.0655	0.0595	0.1045
Heat transfer rate U	16.8	13.89	16.8	15.26	16.8	9.57

Notes: The tube-side fouling resistances are all shown corrected by an A_o/A_i factor of 7.5. Thus, the tube-side fouling resistance for case 1 is 0.001, and that for case 3 is 0.006. Unadjusted total fouling resistance (shell side plus tube side) for all three cases is 0.006.

the inside of finned tubes increases the surface required. On the other hand, the effect of fouling on the fin side is minimal. Consider only case 3 of Table 10-4. The shell side rate and fouling resistances total 0.030. The tube side heat transfer and fouling resistances are (0.001 + 0.006) or 0.007. This value, corrected to the outside surface, is (0.007 × 7.5) or 0.0525 which is greater than the shell side resistance, (0.030). As tube side resistances approach the value of outside resistances, the benefit of fins decreases. This selection only is in the grey area where another design (lower fin height or lowfins) is apt to be more economically feasible.

Raters should take particular note of the following. The overall rate, U, and the surface of the exchanger, A, are essentially inversely proportional to one another. In these examples the increase in surface to account for fouling in case 1 is (16.28/13.89) or 1.2095 or 20.95%; for case 2 it is (16.28/15.26) or 1.10 or 10%; in case 3, it is (16.28/9.57) or 1.755 or 75.5%. From these numbers it can be said that in double-pipe exchangers the effect of fouling on the exchanger's size is often less than for other designs. The exception is case 3 where the benefit of fins decreases for reasons noted above.

10.2.4 Air Cooler Fouling

The effect of tube-side fouling on the selection of an air cooler can be understood by comparing the tube-side resistances of several tube-side fluids. Table 10-5 shows resistances for six fluids or types of fluids, without accounting for tube-side fouling resistance.

In cases 1 and 2, the air-side resistance (0.10) is substantially more than the tube-side resistance (0.0005 or 0.001). In these applications the use of high fins is beneficial. In cases 3 and 4, the heat transfer resistances of light hydrocarbons and lube oils are respectively 33 and 14 times (0.003 versus 0.100 and 0.007 versus 0.100) less than air. High fins remain beneficial. In cases 5 and 6, the heat transfer resistances of heavier oils and asphalts are coming nearer to that of air, and fins are less beneficial. For these conditions it is usually best to use lower-height fins. Depending on the tube-side pressure drop, available alternate choices include larger-diameter tubes or tubes

Table 10-5 Typical Tube- and Air-Side Resistances in Air Coolers, h · sq ft · °F/Btu

Case	Typical fluid(s)	Typical air cooler air-side resistance	Typical tube-side resistance based on inside surface	Tube-side resistance $r \times A_o/A_i$ of 30
1.	Steam or hot water	0.10	0.0005	0.015
2.	Cold water	0.10	0.001	0.030
3.	Light hydrocarbons	0.10	0.003	0.090
4.	Lube oil	0.10	0.007	0.210
5.	Heavier oils	0.10	0.020	0.600
6.	Asphalt	0.10	0.040	1.20

without fins. The latter are called bare-tube sections. These trial-and-error efforts are a variation of the iterative process. When sizing air coolers, the tube-side heat transfer resistance and fouling are additive. If a 0.002 fouling factor were added to the 0.0005 steam resistance, the required exchanger to provide cooling would be much larger. The same fouling factor added to asphalt (0.040 + 0.002) has a much reduced effect on the size of the exchanger. Tube-side fouling strongly affects the size of the exchanger, as illustrated here. This effort is part of the iterative process.

10.3 DEALING WITH THE CONSEQUENCES OF FOULING

The economic effects of dealing with fouling vary from one type of plant or application to another. Some applications require plants to be shut down for cleaning. In a refinery, that means loss of production; in a power plant, loss of salable power; and in a chemical plant, loss of product or production. A/C systems, on the other hand, may be cleaned at night or on weekends when the unit is turned off, with no adverse effect. In other applications, the consequences depend on specific requirements or conditions.

10.4 WAYS OF MINIMIZING THE EFFECT OF FOULING

Fouling should be addressed so that exchangers do not have to be shut down often or for long periods for cleaning. Many steps can be taken to keep systems on-line. In general, users can find ways to reduce or prevent fouling in the first place, to reduce the severity of fouling when it occurs, or to make cleaning less disruptive when it has to be done. These goals can be approached on two levels: choice of equipment or construction, and attention to the water stream.

The equipment planning stage offers many opportunities for avoiding economic consequences of fouling. The plant's history can help the engineer determine a fouling value for sizing purposes. Adding surface to an exchanger can, in effect, increase the fouling factor. Installing automatic brush cleaners can reduce fouling as it occurs (see Section 14.5). Any fouling stream should be kept out of the shell of a shell-and-tube exchanger unless cleaning lanes are provided. Certain types of construction can simplify cleaning, such as removing plugs from shoulder-plug headers or designing cover-plate headers to have lightweight covers.

There are direct and indirect ways of treating the stream to reduce fouling. The most direct methods are chemically treating the water (although allowance has to be made for the effect on materials and gaskets) and adding strainers or filters to remove impurities. Backwashing will reduce fouling, and increasing the water velocity can provide some scouring action.

In instances when it is most urgent to have an exchanger on-line, a second exchanger can be added in parallel so that the primary exchanger can be cleaned while the plant keeps operating.

10.5 ACCOUNTING FOR FOULING WHEN SELECTING AN EXCHANGER

Designing for too small a fouling factor may result in the selection of an undersized exchanger. However, there are often ways of increasing an exchanger's surface or otherwise improving its fouling factor. When choosing plate exchangers, select designs whose surface can be increased by adding plates. This decision should be made before a selection is finalized. Confirm that nozzles will be large enough to handle an increase in flow. On double pipe-exchangers, replace a steel tube-steel fin element with a steel tube that has copper or aluminum fins. Either change can improve the overall heat transfer rate, allowing for the use of a smaller exchanger. This effectively increases the fouling factor (Section 7.1).

Cleaning requirements can be an important factor in selecting an exchanger. Various exchanger designs involve specific methods of cleaning:

- The tube side of shell-and-tube exchangers can be mechanically cleaned if the proper head design is used. If mechanical cleaning of the shell side is required, cleaning lanes (square or rotated square pitch) should be provided.
- Both sides of plate exchangers can be cleaned mechanically. If cleaning both streams is a requirement, this is excellent reason for selecting this design.
- Both sides of double-pipe exchangers can be cleaned mechanically. The same tubes, installed in shell-and-tube exchangers, cannot be cleaned as readily because they are not removable.
- In air coolers the number of fouling streams is reduced to one, the tube side. This stream can be cleaned mechanically with the proper closure.

The effect of fouling depends on where the streams are placed. For example, fouling has little effect on the fin side of longitudinal-fin tubes. The effect of tube-side fouling on the overall size of the exchanger is largely due to the A_o/A_i factor.

A recurring problem is deciding what fouling factor to design for when the user does not define the requirement. The fouling factors recommended by the Tubular Exchanger Manufacturers Association (TEMA) are a good starting point; they are given in Table 10-1. But before deciding the amount of fouling allowance to include on your project (not all designs are to TEMA standards), you should consider several factors:

1. First, there are cost trade-offs in designing for a high or a low fouling factor. Lower fouling factor allowances can lead to a choice of smaller, less expensive exchangers. What is the cost of an exchanger with sufficient fouling allowance? What is the possibility of a plant shutdown due to too small a choice of exchanger? How much is saved by selecting a unit with less fouling protection? Are the savings worth the risk?
2. Adding surface to deal with fouling is also a complex issue. There is always a possibility that an undersized exchanger will be furnished. To

be prepared, determine how surface could be added and if space is available to make this change. For example, if an exchanger has a 6-in-diameter shell with 4-ft tubes, increasing the length of tubes to 6 ft increases the surface 50% for a small increase in cost that could minimize the fouling problem. However, this is not likely to be a good solution for large exchangers unless the increased tube length is manageable.

3. Another factor is turnaround time. Some units require substantially more time to clean than others. The question is, Will the unit meet time constraints?
4. What is the quality of the available water? Is it cruddy or of high hardness, or does it have other properties conducive to fouling? Can the exchanger be cleaned by backwashing or by the use of brush cleaners (Section 14.5)?

Keep in mind that fouling does not affect all exchanger designs equally. A fouling factor of 0.002 (hr °F sq ft / BTU) on the viscous side of a longitudinally finned exchanger has almost no effect on the exchanger size, but the same fouling factor found inside tubes with external fins has a major impact. High fouling allowance affects the size of plate exchangers. Selecting the best fouling factor to use is not easy. One water-cooled unit sized with a fouling factor of 0.002 (hr °F sq ft / BTU) underperformed after three weeks in service; Several applications designed for fouling factors of 0.001 (hr °F sq ft / BTU) were clean after many years of service. Thought must be given to the amount of fouling to be allowed for in each stream before sizing begins.

Chapter 11
Manifolds and Headers

Manifolds, usually circular in cross-section, are devices intended to distribute flow as evenly as possible into two or more streams in parallel. Assume a selection has four exchangers in parallel. In this case the goal is to design a manifold such that one-fourth of the flow goes to each exchanger which would be good distribution. Poorly designed manifolds will not meet this goal. Heat exchanger headers (rectangular in cross-section or modified with rounded corners) act like manifolds as their purpose is to distribute flow equally to tubes in parallel. Poor distribution has been the cause of trouble in both manifold and header designs. These subjects are addressed in this chapter. In heat exchanger applications the term *manifold* refers to a piping arrangement designed to distribute flow to exchangers or sections, but not within headers. Manifolds are not part of the exchanger, nor are they normally furnished by the heat exchanger manufacturer. This said, simplifying their design and supports are part of the exchanger selection process. Headers and manifolds should be finalized together. When poorly designed, the exchanger may not and often does not deliver its full heating or cooling potential as will be illustrated by examples. Manifold and header designs are a governing conditions in the selection of air coolers.

11.1 SELECTING MANIFOLDS

Well-designed manifolds will provide the flow distribution needed to assure a unit's performance. Sometimes manifolds cannot be designed to meet this criteria. When this occurs it is probably best to change the exchanger selection. An example follows.

Suppose the best selection based on heat transfer needs and low cost surface is seven 9-ft 0-in sections in parallel (width 63 ft). There isn't a good way of building a manifold to distribute flow equally to seven streams in parallel. Here are suggestions to simplify piping by changing the selection. An alternative might be eight 8-ft 0-in wide sections (64 ft total width) arranged 2 streams of 4 in parallel or 4 streams of 2 in parallel. A manifold of 2 streams of 4 units in parallel is shown in Figure 11-1. Another choice might be six 10-ft 8-in wide sections (64 ft width) arranged in 2 streams of 3 in parallel or 3 streams of 2 in parallel. The goal is that every stream in parallel and to every section and connection have the same length, size of piping, pressure drop, and carry the same flow. In other words the pressure loss through each piping path should be the same. Inexpensive manifolds could be built that do not meet this criteria. If manifolds are left for others to decide, the engineer

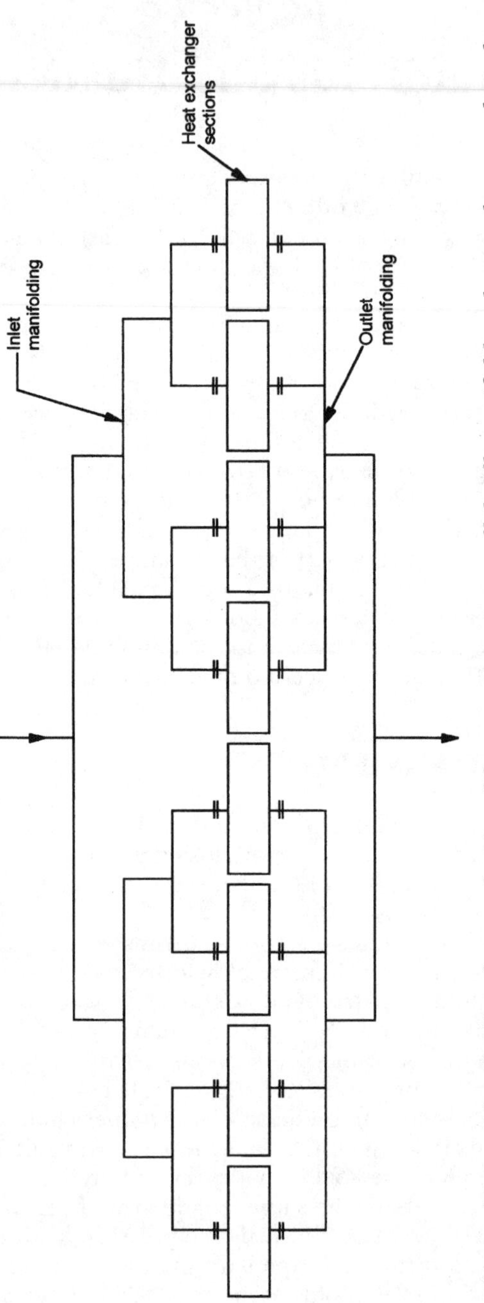

Figure 11-1. Manifolding for two banks of four sections in parallel. All manifold paths are the same lengths. The outlet manifold is not as complicated, particularly if the leaving fluid is all liquid.

is apt to be asked why a unit is not meeting its performance guarantee. Poor distribution or not accounting for manifold and header expansions and contractions are often the reason. Why not address these before they become an issue?

11.2 AIR COOLER HEADERS AND THEIR LIMITATIONS

Shell-and tube units can have large inlet and outlet nozzles (12 in, 14 in, 16 in, 24 in, etc.) built on the exchanger. Plate exchangers can also have large diameter nozzles but their limits are usually on the low side of this listing. Check with the manufacturer for available sizes.

Air cooler headers are basically flat plates. Flat plates cannot sustain pressures anywhere near that of cylindrical shapes. One way of increasing their ability to contain pressure is to reduce their effective size, particularly by stiffening their flat surfaces using welded-in pass partition plates, or minimizing the width of top and bottom header plates. Manufacturers often do this. Header widths, due to pressure constraints, are usually limited in width to 6 or 8 in nozzles. When air coolers require a large supply nozzle, it is common to supply a manifold to mate with it that has several take-offs. These take-offs are limited by the nozzle size that can be connected to the header.

In sizing any exchanger but particularly air coolers because of their manifold requirements, a rule of thumb independent of manifolding proven to be effective is that the pressure drop through the tubes should be at least 5 times larger than the inlet and outlet pressure losses combined to have good distribution in the tubes. The cross-sectional areas of headers is limited. As headers become larger, the allowable design pressure they can sustain drops. A relative idea of the magnitude of this drop can be noted by referring to Table 4-2 and observing the difference in design pressure for four-, five-, and six-layer sections. This is another way of saying that the unsupported header dimensions are further apart. To offset his it is usually economically sound to increase the number of units in parallel. This makes more positions available for nozzles and aids in good distribution. When straight-travel and entrance pressure losses are about the same in value, good distribution usually does not happen.

Table 4-2 shows the difference in design pressure for four-, five-, and six-layer sections. Consider the header's top and bottom plates. These are width limited by design pressure just as the vertical sides are. Usually their limiting width is that of an inlet nozzle, 6 or 8 in. One way of reducing this dimension is to take a given nozzle size and flatten the inlet where it will be welded to the header creating an oval shape that is the width of the header. This will reduce the nozzle's entrance a small amount but allow for the use of a smaller header.

The goal is to avoid trouble jobs. Header designs are sometimes the cause. Figure 19.6 shows the cross-section of a header and welded-in pass-partition plate. Assume the 36 tubes 1-in 12 BWG in this layer are the first pass. The header cross-section is 7 in wide and the distance above the pass partition plate is 3 in. The following applies:

- The first pass flow area is 7 in × 3 in (neglecting the area occupied by rounded corners of the header) or 21.0 sq in.
- The flow area of the 6-in Sch 80 inlet nozzle is 26.07 sq in.
- The flow area through (36) 1-in 12 BWG tubes is (36) (0.479) = 17.24 sq in.

Consider a particle of fluid passing through the unit. It passes the 26.07 sq in nozzle, the 42 sq in header path (the flow divided at this juncture), and the 17.24 sq in of tubing. These are relatively equal areas and the unit will probably perform at these conditions (it also depends on the flow rate).

Several trouble jobs occurred as a result of manufacturer's cost reduction drives. Here is an example that has occurred more than once.

A 6-in nozzle is flattened into an oval shape 4 in wide and welded to a header of the same width. The decision was that a header made by welding four plates together would be less expensive than welding bent shapes as shown in Figure 19-6. In the rectangular header, the first pass flow area is 4 in × 1-1/4 in or 5 sq in (it was 21.0 sq in using bent plates). The area available as a result of the curvature of the bent header no longer exists. The 1st pass flow area using welded plates is less than 25% of that using bent plates increasing the velocity in the header more than four times. This component of pressure drop is increased 16 times. A flow area of 5 sq in (10 when divided flow is considered) is too small for feeding 17.24 sq in of tubing distorting the flow pattern to each tube.

Knowledge of how to alter the relative areas of headers and nozzles when trying to meet pressure drop needs is frequently useful in applications having low pressure gases, vapors, or a mixture of liquids and vapors. If the calculated pressure drop is too high, it can be reduced in these ways.

1. Use the largest standard cross-sectional headers available. Avoid the smaller headers of liquid coolers.
2. Increase the entering nozzles sizes. Consider the use of nozzles flattened so that their entering shape is oval.
3. Add a third nozzle on the center line of the header.
4. Change the tubing size to one having a larger diameter. Make sure a tube support exists for the new design. Note: A 1-in OD tube with 5/8-in hi fins has the same OD as a 1-1/4 in tube with 1/2-in hi fins. Thus, a tube support design exists.

11.3 HEADER NOZZLE LOCATIONS AND QUARTER POINTS

The location of a nozzle or nozzles affects distribution and may affect performance as well. A lone inlet nozzle is generally positioned on the center line of a header. In this arrangement half the flow will be to the left and half to the right. When two inlets are installed on a header, the nozzles are generally positioned at the quarter points. The reason is that after leaving the inlet nozzle the flow will divide with half going to the left and half to the right. The lesson to be learned from the example that follows in Section 11-4 is, Do not relocate nozzles until convinced that the change satisfies distribution needs. Inlet nozzles and their locations determine flow distribution in tubes more than outlet nozzles.

11.4 LUBE OIL COOLER HEADER DESIGN

In one instance an order was placed for many coolers. The size needed was halfway between two standard catalog items. Because of quantity and the need for low price, it was decided to build units that exactly met the client's need. A new header with a cross-sectional flow area that exactly met the operating condition with no excess capacity was designed for the application. A change, requested by the user, resulted in the following problem.

In the as-built design the inlet nozzle was positioned on the centerline of the header. Half the flow was to the left of the centerline and half to the right. The unit was tested and met the performance guarantee. After one unit was delivered, the client asked that all following units have the inlet nozzle relocated to one end of the header to simplify their piping, see Figure 11-2. The request was granted without engineering approval, a frequent cause of problems. This particular change would not be a problem in most cases. Here it was. All flow in the header was now in one direction only with its velocity doubled and local pressure drop increased by four. The header was now undersized causing a reduced and uneven flow distribution to the tubes. The best estimate of the as-built unit was that three-fourths of the flow traveled through one-fourth of the tubes, resulting in a 70% drop in the desired cooling.

Whenever this kind of problem occurs the engineer's goal can be any or all of the following: arrive at the best fix, meet a delivery requirement, keep cost to a minimum, or get a plant back on stream quickly. The corrective measure begins with the knowledge that the unit performed before the nozzle was relocated. The cause of non-performance was that the flow area in the header was reduced by half effectively doubling the velocity and increasing the pressure loss by four at the inlet. Too small a supply flow area to too many tubes leads to uneven flow to tubes and a resulting drop in performance. The fix agreed to was to add another nozzle in the header and leave the end nozzle in place. This adds two new areas for flow, one in each direction from the new nozzle. In effect this triples the header area from that existing before this change is made. The fix requires a new take-off from the supply line dividing flow so that it enters the exchanger at two locations. This was simpler than removing the nozzle and sealing the header. It also minimized delays. After the change was made, the units performed as designed. Had the headers not been built when the problem surfaced, a better fix would have been to build larger headers initially and locate the inlet at the end desired. In effect the parallel supply manifold is equivalent to enlarging the header.

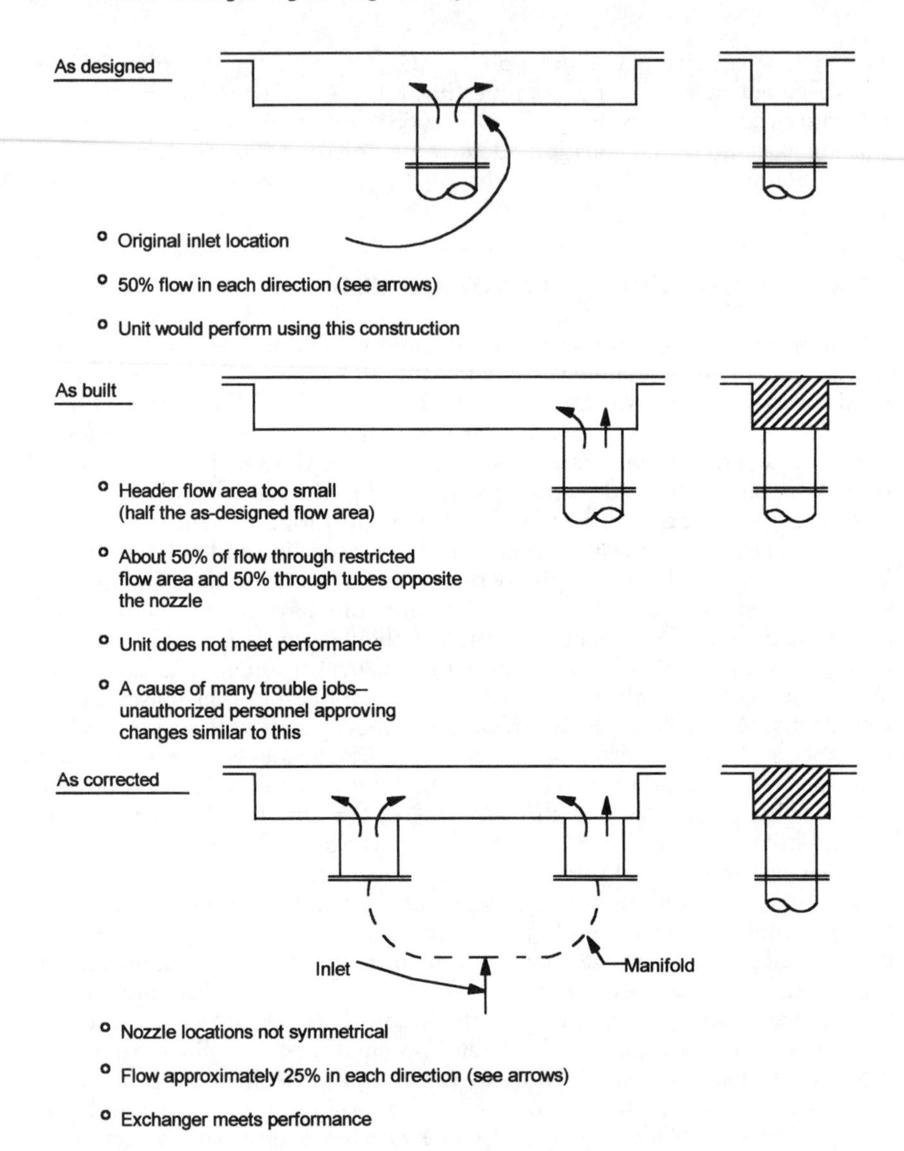

Figure 11-2. Lube oil cooler distribution problem.

11.5 FOUR LIQUID COOLERS IN PARALLEL

One reason why manifolds may leak is thermal gradients. Here is an example. Consider four liquid coolers in parallel that have manifolds as shown in Figure 11-3. In this arrangement, flow through the individual coolers is not uniform, because the entering pressure loss at the separate coolers is not the

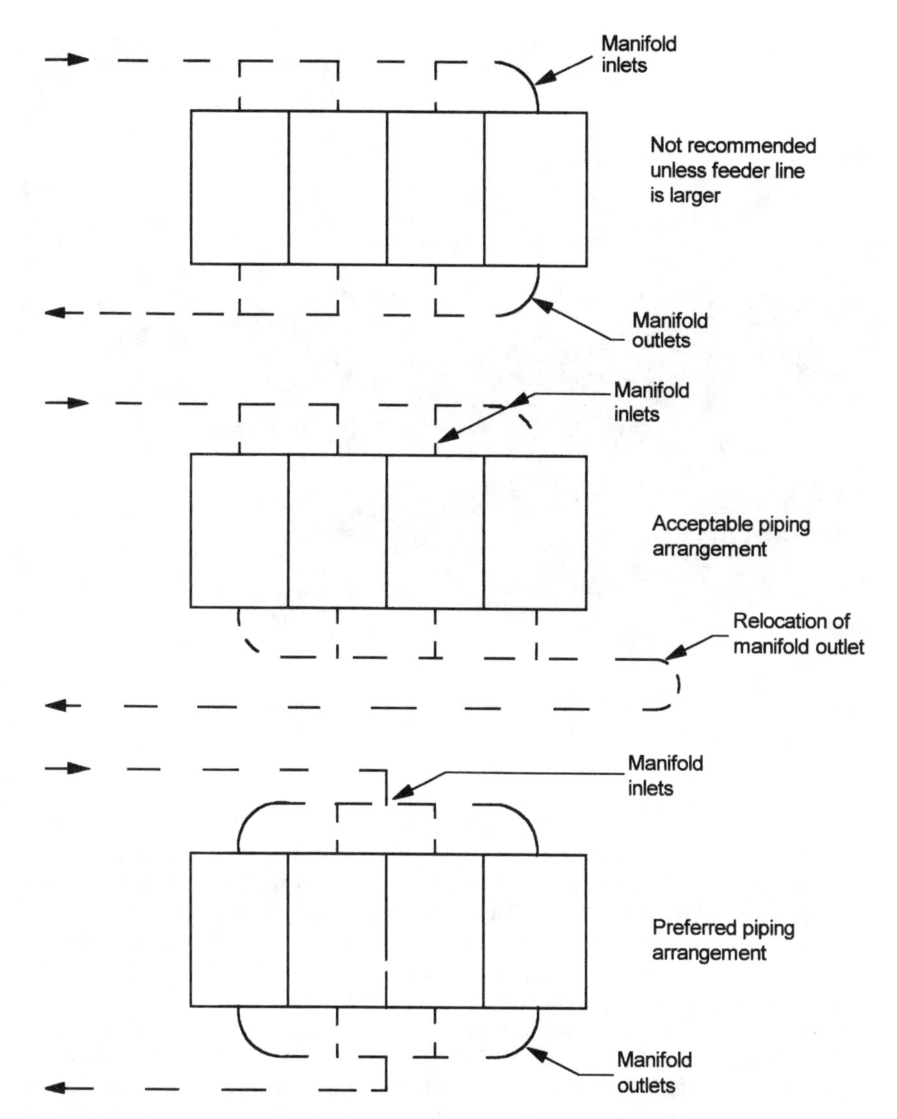

Figure 11-3. Lube oil cooler manifolding, shell side: double-pipe exchangers in parallel.

same. At the first take-off, 100% of flow passes; at the second, about 75% passes; at the third branch, about 50%; and the remainder passes at the last nozzle. The inlet pressure loss is different at each location: the velocity-squared values differ. Since these are not equal, the flow to the coolers is not equal. There will be a manifold differential expansion at the outlets because

Figure 11-4. Bank of double-pipe exchangers and manifold. (Reprinted by permission of Brown Fintube Company.)

of slightly different temperatures there. The coolers' nozzles are fixed in position, while the manifold will expand. Eventually this leads to nozzle faces sliding relative to each other, with leaks developing at the gasketed faces. One recommended fix is to design the manifolding so that flow to each exchanger is more evenly distributed (see Figure 11-3) thus minimizing the chances that thermal expansion will be a problem.

Another solution is to moderately oversize the entering manifold. This applies to both streams. An example is shown for the tube side of a bank of double-pipe exchangers (see Figure 11-4). The larger diameter manifold reduces the entering velocity head to all units, thus making the length of flow pressure-drop component large in comparison with the entering pressure drop. This is an excellent way of balancing flow in exchangers. The result will usually be similar outlet temperatures at all units.

Chapter 12
Freezing

Exchanger selections should be made on the basis of governing conditions. Conditions overlooked in sizing an exchanger are frequently the cause of failures. One of the most common results from the freezing of water. Water expands when changing from liquid to ice. Strong forces are generated in this process, enough to crack tubes. Typically, only a few tubes crack, but this is enough, for example, to make a plate-fin bundle worthless, as the tubes are not replaceable. Problems such as this can be avoided if they are addressed at the time the exchanger is selected.

This chapter illustrates some steps that have proved to be beneficial in preventing freezing. Which one is best generally depends on the application.

12.1 ANTIFREEZE

Adding ethylene-glycol (antifreeze) to water lowers the freezing temperature of the mixture. The more antifreeze added, on a percentage basis, the lower will be the freeze temperature. The lowest temperature at which freezing can be avoided occurs when the mixture is near 70% glycol by weight. Any further increase in the percentage of ethylene-glycol will raise the temperature at which freezing occurs until the solution is 100% ethylene-glycol at which point the freeze temperature will be above that of water.

Ethylene-glycol is expensive when compared with water. Should a large quantity be needed, choosing antifreeze to avoid freezing is not a reasonable solution. The author is unaware of any application where antifreeze was used other than in a closed system. Let us look at an instance where the choice of antifreeze created a problem.

Two common glycol-water solutions are 32% and 55% by weight glycol. The 55% by weight glycol solution is usually easier to obtain and provides freeze protection to a lower temperature than the 32% solution. It is more viscous at lower temperatures. Maintenance personnel have a tendency to use these fluids interchangeably which in most cases is reasonable.

The power output of gas turbines and the lubricating oil cooling load increase with decreasing ambient. System heat exchangers will deliver the increased cooling load that occurs with decreasing temperature because the ambient is lower resulting in a larger LMTD. However, a limiting condition occurs near 10°F for the 55% solution and 0°F for the 32% solution. At these points the exchanger's cooling capacity decreases substantially due to flow in the exchanger reaching the transition region. A trade-off ensues. Which is more important, better freeze protection (55% glycol solution) or the 10°F colder condition that results in more power output (32% glycol solution). If the goal is more power and freezing is avoided the 32% solution should be chosen. A problem that occurs following installation is that maintenance personnel will use a 55% solution in the system because that is what is available. This will cause a reduction in power output. In nearly all other applications, the substitution of a 55% solution would not be a problem. When a system's power output declines look for a change in the percentage of glycol present as a possible cause.

12.2 DRAINING

Another method of preventing freezing is to drain the unit when it is not in use. If this procedure is to be followed make certain that the need for drains is given in the specifications. This solution is ideal if the unit is used in the summer only, which is not the usual case. The drawback is that it takes time to drain and refill a unit. Water chemistry must be checked after each refilling, which costs money. The potential freeze condition could recur several days in succession; it makes no sense to drain and refill the unit each time. Units with more than two passes may be difficult to drain. Internal passes have pockets where water collects and will not drain. Weep holes are sometimes necessary to assure complete drainage. Drilling is easy during manufacture but difficult in the field due to access constraints.

However, there are times when draining a unit is the best way to prevent freezing. These include times when a system is off-line for maintenance or a plant is down for a long week-end. Draining a unit can be avoided by providing some form of heat to prevent freezing.

12.3 HEATING AIR OR WATER

Freezing can be prevented by adding heat to air or water. Warm air can prevent water in the tubes from freezing. Heating water above 32°F prevents it from freezing. The needed heating can be accomplished in several ways:

1. Recirculation (see Section 20.4)
2. Electrically—for example, by wrapping lines with heating cables or tapes.
3. Steam coils
4. Hot water coils
5. Parallel flow (see Section 2.3)
6. Fuel heaters

12.4 AIR RECIRCULATION IN DUCT SYSTEMS

The duct system should be designed to accommodate air being recirculated from a room or building with make-up air. This system is not part of the HVAC unit, but if the ducting is not properly designed, water can freeze on the inside or outside of the coil of an air-handling unit. Some causes of freezing are listed here.

1. A poor design will not provide a good mix of recirculating and make-up air streams (see Section 20.4, Duct Design).
2. The control sensor(s) do not activate the heater(s) when needed (see Section 12.5).
3. Do not overlook week-end, shutdown, or other times when the recirculating blower is apt to be "off". Winds can blow cold air into the make-up air duct where it can travel to the AHU and cause freezing. Do not assume this will not be a problem just because the AHU is located in a warm area and a good distance from the outside. Remember, the AHU contacts cold air in the duct, not the warm air outside it.
4. Undersized recirculating or make-up air ducts or louvers that partially block flow can distort the quantities of air being mixed.

12.5 ELECTRIC HEATERS

For energy efficiency some room or building air is recirculated. This recirculating air stream mixes with incoming fresh make-up air. In cold weather the entering air can cause water in air handling units to freeze on the inside of tubes causing coil failure, or it can freeze air moisture on the outside of tubes provided the fluid in the tubes is below freezing. This latter condition usually does not cause coil failure.

One way to avoid freeze conditions is to install electric heaters to warm the air to a temperature that is above freezing. This design is commonly used and performs well in most applications. However, failures have occurred for the following reasons.

1. Insufficient space is provided for the streams to completely mix resulting in the coil freezing (Section 20-4).
2. Temperature sensors located in the space where the streams mix may not provide a reliable control temperature that will turn on a heater or control a mix temperature. In one case the author measured entering temperatures at several areas near the coil face. These ranged from 68°F at one location to 28°F at another. The latter temperature was taken in the area where the coil failed.
3. Consider a thermostat installed in a room for the purpose of controlling room temperature. If at some time during the day the sun shines directly on the thermostat, the sun's energy will warm the sensor and in effect raise the control point thus preventing the heater from being activated. Local conditions in ducts, analogous to this, often occur preventing heaters from coming on when needed. The problem the engineer faces is locating the sensor(s) so the heater(s) is activated when needed.

Some installations have sensors in both recirculation and makeup air ducts. Problems have occurred when one activates a heater while the other is off, depending on conditions at either sensor's location. When the makeup air heater is off, the potential for freezing or icing in cold weather increases. In other words, thought must be given to the control sequence and sensor location.

Control points should not be altered at a future date without confirming that the sequence for activating heaters is maintained. In the control sequence, a prime consideration should be, What happens on cold weekends when a recirculating blower may be off? During this time coils frequently fail as cold air can reach the coil due to outside winds. One cause may be that maintenance personnel often have the weekend off and are not available to react. Keep in mind that replacing a coil will get the system back on-line but does not address the problem that caused the failure. If it is the only corrective step taken, the coil is apt to fail again.

12.6 HOT WATER AND STEAM COILS

In refinery and chemical plants excess steam or hot water is usually available for heating. Either fluid can be used to heat the air going to air coolers. Steam coils are usually a single layer of tubes spaced on twice the pitch of the process coil. Consider the case of using 250°F steam to heat air. The fin temperature will be near 245°F in all cases. At this temperature air is heated well above freezing, as high as 200°F. This heats nearby air, exactly what is wanted. The tube metal temperature will be less if steam enters at a lower pressure but generally will be high enough to use the one-row coil. When water above 100°F is in the tubes rather than steam, it usually provides sufficient heat to prevent freezing.

12.7 OTHER CAUSES OF FREEZING, AND OTHER PRECAUTIONS

In addition to reasons already given in this chapter for freezing failure of an exchanger coil, the following possibilities should be addressed in preparing specifications:

- If a heater fails, it will not be available when needed.
- A heater will not be on during a power outage.
- Overcooling (freezing) can occur because the heat transfer rate U was too high when the coil was clean.

A parallel flow design is sometimes used to prevent freezing and it is commonly used to prevent channeling, as discussed in Section 2.3. Heaters are often helpful in conjunction with the parallel flow design to warm air or product at start-up, shutdown, or low flow conditions.

Industrial heaters are available that heat air to prevent freezing. These operate on principles similar to household oil-fired heaters. Make sure that the fuel used will flow by gravity to the burner when ambient temperatures are low.

Chapter 13

The Iterative Process of Exchanger Selection

After an initial, provisional selection the rater has an idea of the size of the exchanger needed. The first selection is not necessarily the best choice. Several iterations are usually needed to reach this goal. The number of iterations often increases with exchanger size. The engineer should decide if the provisional selection is competitive and has a reasonable chance of being selected for the application. If not, will varying the design make this selection more feasible? A "no" answer to both questions should stop work on the selection. No exchanger is suitable for all applications. To make selections and prepare quotations takes time and money.

Several designs can be named that are rarely used in specific applications. Double-pipe and plate exchangers are seldom placed in service on over-the-road vehicles because of their length and piping needs. Wide exchangers may be competitive if shipped by barge. Code-constructed shell-and-tube units smaller than 15 inches in diameter in liquid service are seldom competitive with double-pipe exchangers. If you are aware that a product is competitive, work should proceed. Otherwise, it should cease. Conclusion: Knowing the limits of your chosen product and those of competing products will make this decision easier.

As an example of another limitation, suppose a design calls for a 36-in-diameter shell-and-tube unit. This should not be proposed if the plant's manufacturing and lift limit is a 30-in-diameter shell. Larger units that exceed a plant's limits have been assembled on railroad cars or on open-bed trucks provided that the largest component does not exceed the plant's lift capacity. This challenge should be addressed before proceeding.

The engineer should know how to balance thermal equations. Commonly recurring problems are cited in paragraphs that follow. Sizing is the same whether the construction is lightweight, commercial, to TEMA C, B, or A standards, to API 660 or API 661 standards, to the ASME Code, Section III or VIII, or to power plant codes. Exchangers built to a particular standard vary in cost versus exchangers built to other standards. Thought should be given to the quality of construction needed, as less demanding requirements often suffice and may improve delivery.

For a given application many exchangers may be thermally correct but only one is apt to be the best choice. It might be selected from several shell-and-tube offerings, double-pipe selections, or plate exchangers. The selection process is rarely complete unless economics are part of the decision.

Cost can sometimes be minimized by selecting an alternative unit, though sometimes no realistic alternative exists. A frequent reason for overspending is overspecification in construction or choice of tubing. Water-to-water exchangers cost more for refinery, chemical, or power plant applications because they entail TEMA R or ASME Code construction. Higher-quality construction is used to protect against failures that could cause a plant shutdown. Hence, more demanding construction is often necessary. But most water-to-water exchangers do not require this quality of construction. Commercially constructed exchangers with thinner tubes and tubesheets are available on the market. Rather than two grooves at each rolled tube-to-tubesheet connection, they might have one or none. The packed-head design is common in this service (TEMA type BEP; refer to Figure 3-1). These designs perform well. The benefit is that they are low in cost.

Shell-and-tube exchanger baffles are usually spaced 6 to 30 in apart. One of two exchangers with the same diameter shell and tube length could have up to 5 times (30/6) the number of baffles as the other(s). These add to cost. The same is true of tube-side pass partitions. In both cases making another selection may be best from an economic point of view. More passes mean tubes must be dropped because room is needed for pass partition plates. If more surface is needed, the shell OD may need to be increased, thus adding to cost. (Note: This is the weak point of rating services, which rarely have the means to factor economics into their selections.)

Air coolers differ from shell-and-tube units in terms of cost. They do not have shell side baffles, and their tube supports are independent of tube passes. In most cases tubes are spaced far enough apart that none are dropped to make room for partition plates. The variable is the number of pass partitions and, in air coolers, these have a small effect on cost.

There are few cost variables for double-pipe exchangers. Some have manifolds with many connections. Technically, these are not part of the heat exchanger except that they affect system cost. Plate exchangers do not require shell-side baffles or tube-side pass partitions. They can be selected using a minimum number of units in parallel. Alternate selections are not as common using this design.

The remainder of this chapter details the modifications that can be made to selections of various types of exchangers. Keep in mind that the overall heat transfer rate U is a combination of outside and inside surface heat transfer rates, outside and inside surface fouling, and metal resistance. It is possible to modify these to advantage. The most obvious iteration is to make maximum use of the available pressure drop. Other ways are given in the following sections of the chapter, which were prepared to serve as a basis for checklists.

There is an exception to the recommendations. If the straight-travel pressure drop is not at least 5 times the nozzle and turning pressure drop losses, poor distribution is apt to occur, a fact known from experience. This condition frequently occurs in selecting exchangers for low-pressure gas or vapor streams with small allowable pressure drops or for undersized headers or nozzles in liquid service.

13.1 SHELL-AND-TUBE UNITS

The rating engineer begins by making a selection of the equipment to offer for an application. Many designs can meet this criteria. However, one choice is usually best in terms of cost, delivery, ease of cleaning and handling, mating with existing equipment, fitting in a given space, and meeting the requirements to name a few. Do not assume that all requirements always apply. Often one is waived because its omission can improve one of the above conditions.

A selection can often be improved by making modifications to it that better fit the requirements. This is called the iterative process. A partial list of modifications that can be made to shell-and-tube exchangers are given in Table 13-1. Not listed are subjects such as corrosion allowance, type of flange faces, location of mounting brackets, tube vibration, painting, code stamp, type of welds, and other extras that are not part of the selection process.

The choice of tube metallurgy in shell-and-tube exchangers is not as easily categorized as are other resistances to heat transfer. The metals to use can be influenced by many factors. Several examples will be given. Tubes are almost always the most expensive component of an exchanger. A reduction in their cost can often be attained by substituting one material for another (Section 15.5). For example, titanium costs more than admiralty metal on a per pound basis. Some titanium alloys have tensile strengths roughly 50% greater than admiralty metal and their density is roughly half as much. The design temperature of titanium tubes can be 200°F higher than admiralty metal and thinner tubes can be used. Combined, titanium tubes are often less expensive than other metals for the same surface.

A potential bare tube metallurgy benefit can be attained by answering the question, Is it advantageous to reduce the amount of surface required by installing an automatic brush cleaning system (Section 14.5)? Small diameter tubes reduce the volume of refrigerant needed (Sections 14.6 & 15.1). This question follows, Will standard tube bends of the metal selected be available? Plate exchangers can be built of many materials and are usually smaller in size than bare tube exchangers (Chapter 8, Sections 14.7 & 15.9.2). Will this design be a better choice than shell-and-tube unit(s) of similar materials? Similar reasoning applies to double pipe exchangers.

For a given application the tubes of shell-and-tube units in series need not have the same metallurgy. The largest pressure drop should be taken in those units requiring tubes of costlier metals (Sections 18.5 & 19.10). Tubes with high thermal coefficients of expansion can be useful if the shell-side inlet is installed beyond the U-bends (Section 18.9). If U-tubes are required, the bending radius may be limited by tube metallurgy. The tube thickness must be sufficient to account for thinning that could lower the design pressure to an unacceptable level (Section 3.3.2). In dual wall tube applications it should be confirmed that the needed materials can be obtained and assembled (Section 15.9.1). In the usual case the stream requiring special metallurgy is placed in the tubes of shell-and-tube exchangers because only the tube side need be built of higher quality materials. If placed on the shell side both

Table 13-1 The Iterative Process for Shell-and-Tube Exchangers

Potential ways of improving a shell-and-tube exchanger selection

Shell-side heat transfer rate
- Change baffle spacing and add or subtract baffles.
- Change the baffle cut opening.
- Change the tube pitch.
- Add fins to the outside of tubes: lowfins; longitudinal fins; spiral or plate fins.
- Drop tubes at the shell inlet or outlet.
- Use more of the allowable pressure drop.
- Use two inlets rather than one (divided flow).
- Change the tube diameter.
- Place the product requiring the expensive material in the tubes.

Shell-side fouling
- Place the high fouling stream in the tubes.
- Use backwashing.
- Change to square pitch (cleaning lanes should be continuous).
- Increase the tube pitch.
- Increase the velocity of the product through the shell.

Tube metallurgy
- See Section 13.1.
- Consider cost, quantity, availability, and delivery time.

Tube-side film resistance
- Install turbulators.
- Shorten the tube length to reduce the *L/d* factor.
- Use internally finned tubes.
- Select a different tube diameter.
- Change the number of passes.
- Use more of the available pressure drop.

Tube-side fouling
- Increase the tube-side velocity.
- Consider an automatic tube cleaning system.
- Treat the water chemically.
- Control the water pH.
- Switch to air cooling (this eliminates one fouling stream and simplifies cleaning in the other).

streams often require the higher grade materials or, as an alternate, thicker lower grade materials. Other examples are given in Section 15.9.

The cost of tubing can often be reduced using extended surface or fins. The material to use depends on the application. In pure cross-flow (See Figures 3-2 and 4-6), the heat transfer capability of air can be increased using copper rather than aluminum fins. This change increases the heat transfer efficiency by 2 to 4% (Table 5-1). In liquid flow aluminum or copper fins will be about 25% more efficient, and sometimes much higher, than if the same number of steel fins were used (Section 7.1 and 16.2). Copper fins are somewhat more efficient than aluminum in liquid service. In many applications the use of steel fins is limited (Section 5.2) but there are many applications where they are beneficial. Included are those with high bare tube metal temperatures that cause heated products to be reduced to simpler compounds (Section 18.1). The temperature in contact with the product can be reduced by taking advantage of the high metal resistance of a steel fin/tube combination. The higher resistance lowers the skin temperature and minimizes the chance of high temperature causing a product breakdown. This condition has occurred in tank heaters where light hydrocarbons are vaporized by temperature alone causing a vapor lock in the suction line. Sometimes, high temperatures cause fin tips to burn. Higher strength fin metals have been designed and used to satisfy this condition. Non-ferrous fins can be used up to 550°F. This limit does not always apply as it also depends on the temperature of the tube side fluid and the fin/tube construction. The fin temperature limit differs for each fin design and material.

Lowfin tubes (Sections 5.3, 14.2, 16.5, and 17.7) differ from those having high fins in that flow is in both longitudinal and cross-flow in shell-and-tube units and the tube/fin construction is homogeneous. Higher height fins are usually in cross-flow service alone. In some applications steel fins are preferred because they can be cleaned using caustic solutions. From the foregoing it can be seen that the selection of the tube or the tube and fin materials requires skills in areas other than metallurgy alone.

A selection should not be finalized until answers similar to the following are obtained; if answers are unsatisfactory, use another selection until requirements are met:

1. Has ability to clean the tube and/or shell side been provided?
2. Will lifting the unit into position be a problem?
3. Are covers easy to handle during servicing?
4. Will flagmen be needed during shipment?
5. Can the equipment selected be delivered within time constraints?

13.2 AIR COOLERS

What air cooler is best for the application? In process work and sometimes for industrial air coolers the selection is an iterative one. Questions similar to the following should be answered before a recommendation is finalized; if any are unanswered, a potential for trouble exists:

- Can the equipment be shipped preassembled? Will its manifolding be symmetrical, and will distribution problems be avoided?
- Must the side channels be galvanized? This must be done before the section is assembled; it cannot be done after.
- Does the equipment size and shape suggest excessive shipping charges? Have problems of shipping, handling, and manifolding been addressed?

- Is the recommended surface adequate for the temperature and pressure specified?
- If future expansion conditions are known, have subjects such as distribution, manifolding, floor space, footing loads, isolation valves, and minimum downtime been addressed?
- Have alternate operating conditions been considered?
- If welded tube-to-tubesheet connections are required, will tubes be spaced far enough apart that the heat of welding does not damage fin ends locally?

Perhaps more significant are the following questions:

- What is the application type? If a jacket water cooler, can plate-fin construction be used?
- Must two grooves be supplied at tube-to-tubesheet joints?
- Can thinner tubes be used?
- Is quick delivery a factor?
- The subjects of design and operating pressure, cleaning, and header construction should be addressed together. The significant question is, Should the header be welded box, cover plate, or manifolds? If a cover-plate design is selected, will it meet pressure requirements? Will the design of the cover facilitate cleaning?

All the foregoing should be considered before finalizing the selection and making a recommendation. Users should keep in mind that features wanted and overlooked in the specifications cannot necessarily be added after the equipment is built. Even a simple request like adding an instrument coupling to a nozzle creates a problem with the validity of an ASME Code stamp if the change is made after the stamp is applied. The lesson is to develop a full set of requirements before an order is placed.

The remainder of this section is centered on the iterative process and suggests modifications that can be made to air cooler units that already meet or are close to meeting the performance requirements. One or several modifications may reduce an exchanger's cost, adapt it to the available space, reduce its noise level, lower the fan HP and cost of operating the unit, simplify its handling, satisfy a delivery requirement, ease mating with existing equipment, or for other reasons. Don't overlook using two fans per bay.

Some ways of modifying air cooler units are given in Table 13-2, "The Iterative Process for Air Coolers." Other possible changes have not been listed because the are usually specific to a given manufacturer. For example, fin manufacturers have different ways of stamping or creasing fin material to create turbulence that can improve the heat transfer rate of an air stream by up to 15%. Another is the way extended surface is manufactured. Some clearance is needed between fins for tools to space the fins and assure good contact is attained between tube and fin. This manufacturing technique limits the number of fins/in that can be applied to a tube. The limits for wrap-on fins for 1-in OD tubes is roughly 6 to 12 fins/in. The fin count will vary with the tube diameter and type of construction.

Table 13-2 The Iterative Process for Air Coolers

Potential ways of improving on an air-cooler selection

Outside-surface heat transfer rate
- Select a different tube diameter; fin height, thickness, or material; number of fins; or number of tubes. Alternatively, choose a bare- or lowfin-tube unit.
- Choose a different number of layers.
- Select a shoulder-plug, cover-plate, or manifold header.
- See that cleaning needs will be satisfied using the design selected.
- Determine that operating and design pressures and temperatures will be met.
- Provide for the minimum number of nozzles.
- Be sure that tube length mates with existing or proposed exchangers.

Outside-surface fouling factors (usually not a consideration)
- Is the air-side fouling minimal?
- Can a blower or steam lance be used for cleaning to minimize fouling?

Modifying metal resistance
- To decrease resistance, use copper or aluminum fins (Section 5.1 and Table 5-1).
- If a higher resistance is beneficial, the metal resistance will increase substantially if the fin material is other than copper or aluminum.
- Change one or more of the following:
  - Tube OD ($^1/_4$, $^5/_{16}$, $^3/_8$, $^1/_2$, $^5/_8$, $^3/_4$, $^7/_8$, 1, or 1$^1/_4$ in, etc.)
  - Fin height ($^1/_8$, $^1/_4$, $^3/_8$, $^1/_2$, $^5/_8$, $^3/_4$; plate fin)
  - Fins per inch (4 to 14)
  - Fin thickness (generally, 0.006 to 0.02 in thick for aluminum or copper; roughly 2 to 3 times greater for other materials, resulting in fewer fins per inch for these)
- Use integral fins.
- Use both internally and externally finned tubes.
- Select a higher-quality and thinner tube with higher tensile strength.
- Consider material cost, availability, and delivery time.
- Nonstandard construction may be beneficial. For example, use lowfins with 30 rather than 16 fins per inch.

Tube-side film resistance
- Use turbulators.
- Shorten the tube length to reduce the *L/d* factor.
- Select a tube diameter that makes better use of pressure drop or space.
- Increase or decrease the number of passes.
- See the discussion in Section 2.4.

Tube-side fouling
- Decrease the OA/IA ratio to minimize the effect of fouling.
- Use an automatic brush cleaner, chemically treat the water, or add filters.
- Determine the effect of fouling on other plant services.
- See Section 10.2.4 for variations in calculations.

Stamped fins (plate fins) are not limited by the space needed for tools; hence, up to 20 fins/in can usually be offered. In the manufacture of lowfins the usual offering has 16 to 19 fins/in. This is the ideal ratio for most applications (Section 5.3, 14.2, and 17.7), which is probably the reason that manufacturers have standardized on this spacing. However, there are applications where lowfins are more beneficial using about 30 fins/in (Section 5.3). Some suggested air cooler modifications for propeller fan air coolers do not always apply to air handling unit coolers because these are limited by duct aspect ratio considerations (Section 6.1). Then too, if the air coolers are rack mounted, structural considerations often dictate the tube length.

13.3 DOUBLE-PIPE EXCHANGERS

The subjects of double-pipe exchangers and longitudinal-finned tubes in shell-and-tube exchangers are addressed together in this section as they usually use the same tube design. Many combinations of tubes and fins can be used in either type of exchanger. Double-pipe units are thought of as liquid-to-liquid exchangers because they are commonly used in these applications. However, such a categorization is oversimplified, when the following is considered.

A liquid exchanger may have an allowable pressure drop of 5 psi, while a spiral- or plate-fin air cooler may have an allowable pressure loss of 0.5 in of water. In one, the low-heat transfer rate liquid is on the longitudinal-fin side, and in the other the low-heat transfer rate fluid, air, is on the spiral-fin side. Low-conductivity fluids are placed on the extended-surface side of exchangers. If allowable pressure losses are compared and stated in the same units, a liquid loss of 5 psi would be 277 times greater than the 0.5 in allowed for air. Low-pressure gases rarely have this large a pressure drop available and are seldom cooled in double-pipe exchangers. Double-pipe units can operate at higher pressure. The length of travel in longitudinally finned 20-ft double-pipe units in series may be 40 or ft 80 or more, while ambient air in air coolers seldom travels more than 6 to 8 in. The length of travel in shell-and-tube units is rarely longer than 20 ft per pass.

Airflow areas are nearly always greater in spiral- or plate-fin sections than through shells that contain longitudinally finned tubes. Gases and vapors almost always have lower densities than liquids, and this has the effect of increasing the difference in available pressure drop. This is a factor when choosing spiral-fin or longitudinal-fin surface. A review of the implied conclusion that longitudinal fins are not beneficial in gas or vapor service is in order for this reason.

The density of gases and vapors increases as pressure increases. This has the effect of reducing pressure drop. Longitudinal-finned tubes, when placed in shell-and-tube units, have a larger shell-side flow area than a comparable number of tubes in double-pipe exchangers arranged in parallel. Shorter-length tubes are seldom economical in double-pipe exchangers but are beneficial in shell-and-tube units. In this service the shell-side flow area is more than double when compared with double-pipe exchanger flow areas, as flow is in one direction only and the length of travel is halved. This arrangement differs from double-pipe units, where the flows in each leg are in opposite directions. This can be compared with bare tubes in a divided-flow shell-and-tube unit—twice the flow area, half the length of travel, and one-eighth the pressure drop. Thus, longitudinal-finned tubes are and have been economical in shell-and-tube units in vapor and gas services, as tank and line heaters, and particularly in high liquid or gas flow situations. Double-pipe and shell-and-tube units can operate at high pressure.

Another consideration follows. Longitudinal fins are welded to tubes or, in an alternate design, placed in grooves so that tube and fins are mechanically bonded to each other. If either design is to be successful, the tube material must be thicker than the bare tubes used in shell-and-tube units. Typically, 2-in tubes have thicknesses of 12, 14, 16, or 18 BWG. The standard double-pipe tube is technically not a tube but a pipe, $1^1/_2$ in Sch. 40. Its average thickness is 0.145 in. Note the distinction of *tube* (2 in OD) and *pipe* ($1^1/_2$ in Sch. 40). One is 2 in OD and the other 1.90 in OD. When bare tubes are used, the added thickness required for finning is not needed. This is a plus if tubes must be of a higher-quality metallurgy. When thinner tubes are used, the tube-side flow area increases greatly.

There are some limits in the use of double-pipe exchangers or longitudinally finned tubes. In the welded fin to tube construction, any material that can be welded to the tube can be used as the fin material. Welding eliminates the use of aluminum or copper fins on steel tubes. If grooved tube construction is used, the fins can be of any material. Keep in mind that just because any material can be placed in the grooves, it does not follow that this is a good thermal choice. Double-pipe construction is rarely used in low pressure gas applications, but shell-and-tube units with spiral fins can be. Another limitation is that pipes are generally thicker than tubes and add to cost.

In the iterative process the following factors apply to longitudinal-fin construction only:

1. Regarding cleaning, the double pipe is ideal if mechanical cleaning is required on both streams. U-bend cleaning is not as limiting in this design as in shell-and-tube units. Similar finned tubes installed in shell-and-tube exchangers cannot normally be mechanically cleaned on the shell side.
2. One inch and larger diameter tubes are commonly used in double-pipe, air cooler, or shell-and-tube exchangers. Their internal flow area is large and these generally do not have a particle size limitation. The fin side flow area of longitudinal fins can be large provided the design offers few fins of high height. As fin height decreases and the number of fins increase, the equivalent diameter decreases. Seldom do particles block passages in standard units though the condition has occurred. In one case flat longitudinal fins were bent into L-shapes. The benefit was seen as packing considerable surface into a small volume. The idea backfired because the small equivalent diameter created blocked flow even though few particles were present. Particle size has been a factor in some turbulator designs that were intended to improve the heat transfer rate. Instead the turbulators and particles combined blocked flow.

Plate exchangers are particle size limited (Section 19.7) as is the shell side of shell-and-tube units built on a tight pitch (Section 15.2). Be aware of these limits when making exchanger recommendations when particles are present.

3. The heat transfer rate in longitudinally finned exchangers can be improved in several ways. A good cut-and-twist design can usually reduce the shell-side resistance in cooling to about 0.28 times the value without it. Metal resistance can be reduced using copper or aluminum fins. This is a plus for embedded fins. If this construction is used, that is, steel tubes and copper or aluminum fins, the exchanger will typically transfer about 25% more heat. (*Note A:* Both of these cooling improvements are offset by a 35% increase, on average, of the μ/μ_w factor.) Finally, steam may be used to heat a product. Heating raises the tube metal temperature and reduces the μ/μ_w factor, reducing the tube-side resistance to about 0.75 times its value (starting-point estimate).
4. Double-pipe exchangers with the same diameter shell are interchangeable in that they can be stacked in series/parallel combinations. This makes them ideal for plant expansions. Tube connectors of various radii are available for this purpose.
5. Large temperature drops are easily handled. For example, a 900°F temperature drop might be spread over four sections in series. Costs can be reduced by using higher-quality metals only in those units that operate at higher temperatures.
6. Design pressure and temperature limits can be high, about 1300°F and 15,000 psi. Few designs perform better at these extremes.
7. The longitudinal fin is ideal for flow in the laminar and transition regions. It is excellent for liquid fluids that have a shell film resistance that is 3 times or more that of the tube fluid.
8. The bare surface in double-pipe exchangers is commonly, but not always, $1^1/_2$-in pipe or 2-in tubes. Therefore, only a few pipes or tubes are needed to carry large flows for long lengths, a characteristic of this construction. As Tables 2-1 and 2-2 indicate, $1^1/_2$- or 2-in diameters allow the use of long-length tubes without large pressure drops. The small number of tubes can mean shells of minimum diameter. The advantage also applies to the shell side of double-pipe exchangers, where pressure drop can again be controlled by selecting the right diameter shell. As shell diameters increase, fin heights must increase to keep tubes concentrically positioned or, alternatively, for bare-tube units, centering devices must be provided.

Another factor to be considered in the iterative process is delivery. The prime requirement may be to get a plant back onstream. In this case a costlier design may be preferable because it best satisfies this need.

Finally, consider the materials needed. Are they available at a mill or warehouse? Some applications require high-quality materials, which can be costly. Generally, materials can be purchased at lower cost when bought in higher quantity. Small quantities can often be purchased from a warehouse at lower cost and for faster delivery. A warehouse may even stock higher-quality metals that could be substituted for the material needed. *Warning:* Never switch materials without the user's approval. (See Section 15.5 regarding substituting tube materials.)

13.4 PLATE EXCHANGERS

A number of iterative options are available in selecting shell-and-tube units, air coolers, and double-pipe exchangers that are not available when selecting plate exchangers. These include tube pitch, type of fin and spacing, use of turbulators, choice of header (cover plate, shoulder plug, and manifold), and external cleaning (steam lance).

However, there are many ways to vary the construction of plate exchangers to improve their heat transfer rate, among them:

1. Selecting the proper spacing between plates
2. Selecting the shape of the herringbone stamping
3. Deciding on the number of plates
4. Making sure cleaning needs are satisfied
5. Shortening the length of travel
6. Changing the material of the plates
7. Refer to Section 10.2.2 for benefits that often apply in competitive situations.

In addition to improving heat transfer rate, one can choose correct nozzle sizes for current or future flows. Nearly all these decisions are made by the manufacturer, who is in the best position to implement changes.

Chapter 14

Techniques to Reduce the Size of an Exchanger

Examples of how to reduce the size of an exchanger follow in this chapter. The ideas given are, for the most part, application-specific. Most of the recommendations depend on the skills of the heat transfer engineer. If the user desires any construction or feature identified, the requirement should be in the exchanger specifications. This is the best way of advising the rating engineer of what is needed. Otherwise, the selection made may not be to your liking.

14.1 COOLING WITH HYDROGEN GAS

The heat transfer rate of a gas is a function of its thermal conductivity and Prandtl number. The heat transfer rate will vary with the gas flowing, but these properties affect the rate more than others. The gas heat transfer rate usually varies by a factor of less than 2 regardless of the gas flowing, with the exception of hydrogen. Pure hydrogen gas may have a transfer rate as much as 7 times greater than air. Exchangers and systems are designed to take advantage of this property of hydrogen. The gas heat transfer rate is part of the overall heat transfer rate, the other factors being fouling, metal resistance, and the heat transfer rate of the fluid flowing in the tubes. If the air-side resistance can be reduced to one-seventh its value using hydrogen gas, the cooler needed will be smaller for the same application. This makes hydrogen gas ideal for cooling of generators.

Hydrogen gas cools generator windings more efficiently than air while using a smaller cooler. Coolers are normally installed in the generator frame, and the smaller size simplifies handling, shipping, and generator assembly problems. The design is much the same as for air coolers.

Sometimes a 90% hydrogen–10% air mixture is preferred over pure hydrogen gas, but its use requires a slightly larger cooler than one using pure hydrogen gas. A larger cooler is required if the mixture is 75% hydrogen–25% air. Keep in mind that air-hydrogen mixtures are a benefit in cooling windings as well as reducing exchanger size.

To understand the sizing of hydrogen coolers it is well to compare its gas properties with that of air. For a given heat load the temperature span of each gas will be the same. The weight of hydrogen gas needed to perform the desired cooling times its specific heat will be equal to the amount of air required times its specific heat.

First, hydrogen gas has a specific heat of 3.5 Btu/lb °F; that of air is 0.24 Btu/lb °F. Since the gas quantities flowing are inversely proportional, the quantity of hydrogen gas needed to deliver the same cooling as air is proportional to the gas specific heats in the same ratio or 6.9% (0.24/3.5) that of air. At this reduced flow, the area needed for hydrogen gas flow is reduced.

For another reason, compare the density of air and hydrogen in terms of molecular weight. Air has a molecular weight of 29, while that of hydrogen gas is 2. Thus the air-side pressure loss should be 14.5 times greater for hydrogen if only density is considered. But pressure drop is also a function of mass velocity squared; for hydrogen versus air it is the reciprocal of $(3.5/0.24)^2$, i.e., 1/212.7, and hence the gas-side pressure drop for hydrogen is much less. For this reason more tube layers which use more pressure drop can be offered for hydrogen gas coolers.

The heat transfer rate using hydrogen is roughly 7 times greater than for air. As this rate becomes closer to that in the tubes, less fin surface is needed. For this reason, it is not uncommon for hydrogen coolers to have fins $^1/_4$ in to $^1/_2$ in high and an OA/IA ratio of between 10 and 15 to 1. Compare this value with that of air coolers, between 13 and 32 to 1. The lower fin height lowers the gas flow area available.

In the building of coolers (as well as other exchangers) there are constraints that have nothing to do with heat transfer. The shape of the cooler must allow for good fan coverage. The shape of process coolers may be limited by their shipping width. The aspect ratio (length to width) of coolers in ducts is normally limited economically to a value of 3; as ducts become wider their side thickness increases, as does their cost, and so available space concerns are almost always present. Many plate-fin coolers are length-limited by finning capabilities.

Hydrogen coolers almost never face these limits. On the other hand, they have this plus: they can be made long and narrow, saving on cost. Remember, however, that for economic reasons, hydrogen gas must be contained to take advantage of its thermal properties.

14.2 LOWFINS

One of the best ways of reducing the size of an exchanger is to replace bare tubes with lowfin tubes. The ends of lowfin tubes are plain and so can be rolled into tubesheets similar to the way that bare tubes are rolled into tubesheets. The lowfin was developed primarily for shell-and-tube exchangers, though lowfins are also used in air coolers. Lowfins are available in many materials including copper, admiralty, red brass, aluminum brass, aluminum bronze, the copper nickels, stainless steels, and carbon steels. Check with manufacturers of these tubes regarding other materials, as many are available provided the quantity required is large enough to cover setup costs. The bare ends of tubes and their fins are the same OD, simplifying assembly in shell-and-tube units.

Lowfins frequently have 16 or 19 fins per inch, which is standard in many applications. Manufacture begins with bare tubes that are between 18 and

12 BWG. The ratio of outside surface to inside surface of the finished tubes will be between 3.2 and 4.8 to 1. Compare this with water resistances that are typically $^{1}/_{3}$ to $^{1}/_{5}$ that of light lubricating oils and whose reciprocals are between 3 and 5 to 1, exactly what is wanted in water-lubricating oil service. Thus this construction is ideal and frequently used in this application. It is also beneficial in other applications. Fin spacings of 16 per inch have been used successfully in liquid-liquid service.

There is no reason why lowfin tubes can't be used in vapor service. They often are. High fin counts are beneficial in medium- and high-pressure superheated steam applications because a higher pressure drop is usually allowed. One advantage is that the tube and fin are integral, making this surface ideal for both services. Fin unraveling cannot occur as with wrap-on fins. Another feature that distinguishes lowfins from other spiral-fin tubes is that in shell-and-tube units tubes are often subjected to both crossflow and longitudinal flow. Most spiral-fin tubes are useful in crossflow service only. Lowfins can be used in many applications.

In the process of extruding fins from the base material, the tube ID is reduced, which has a large effect on tube-side flow, heat transfer rate, and pressure drop. It often happens that as the tube diameter becomes smaller (see Section 2.4, "Tube-Side Considerations in Selecting an Exchanger"), the number of tube-side passes is reduced.

As an example of an application where lowfins are particularly beneficial, assume that a shell-and-tube unit with bare tubes has been performing successfully in a plant for some time. Because of an increased demand for the plant's product, a need exists to expand the plant. It is not uncommon to have a shell-and-tube exchanger's cooling capacity increased by up to 40% by replacing a bare-tube bundle with one that has lowfin tubes. Often no other change needs to be made. This ease of conversion saves on the consequential costs of making changes as it minimizes plant downtime. This bundle could be built while the plant is onstream. The 40% increase in heat flow in one exchanger is about the maximum the author has seen in making this substitution, that is, changing from bare to lowfin tubes. The 40% increase is by no means a limiting value for lowfins. Some applications reduce the number of lowfin tubes required to half that of bare tubes.

For other information on lowfins, see Sections 5.3 and 17.7.

14.3 INTERNALLY AND EXTERNALLY FINNED TUBES

There are applications where the use of tubes having internal and external fins is advantageous. In most of these cases tubes and fins are continuous and of the same material. The outside surface can be greater than the inside surface or vice versa. The surfaces are not likely to be the same shape. These tubes are generally made of a soft metal such as copper, aluminum, or their alloys. The machinery for making them of higher-tensile strength materials such as steel, stainless steel, or low alloy steels is expensive and, in the main, uneconomical. The tubes are often used on repetitive applications such as evaporators and condensers in chillers. This construction is advantageous when tube- and shell-side resistances are close to each other in value. As they diverge, a way to make this construction effective is to increase the outside surface area relative to the inside area or vice versa depending on the magnitude of the resistances. Often, this is easier said than done, particularly if changes are to be made to the inside surface. The tubes are effective where soft metals can be used (lower tensile strength) and the operating pressure does not exceed about 400 psi. Lower-tensile strength materials are of relatively thin construction and for this reason their suitability at higher pressures should be confirmed.

The benefits are best illustrated by example. Suppose a bare-tube unit had the following values, which might be typical for this type of unit:

Shell-side film resistance	0.00125
Metal resistance, copper	0.00003
Tube-side, water resistance	0.0009
Tube fouling, 0.0005 corrected	0.0006
Total resistance R	0.00278
Heat transfer rate U	359

where resistances R are in units of h · ft · °F/Btu and heat transfer rate U is in units of Btu/(h · ft · °F). Now consider a corresponding unit with tubing that has 2 times the external surface and 1.6 times the internal surface of the bare tube. With the same fluid flowing through the tubes of this unit as in the bare-tube unit, the heat transfer rate should be similar to the following:

Shell-side resistance	0.00125	
Shell-side fouling	0.00000	
Metal resistance, copper	0.00007	
Tube-side water resistance	0.00162	
Tube side fouling (0.0005)	0.00090	OA/IA=0.3272/0.1813=1.8050
Total resistance R	0.00384	
Heat transfer rate U	261	

If the bare-tube unit has 400 tubes of $^5/_8$ in OD and are 10 ft long, the gross heat transfer surface will be (400)(0.1636)(10) = 654 ft^2. The product of heat transfer rate and area is then

$$(UA)_{\text{bare tube}} = (359)(654) = 234{,}786$$

A similar calculation for finned tubes in this service, where 2 times the bare-tube surface is assumed, results in

$$(UA)_{\text{finned tube}} = (261)(2)(654) = 341{,}388$$

The number of finned tubes required is (234,786/341,388)400 = 275. The reduction in number of tubes required is 400 – 275, or 125 tubes. Economics must now be addressed. Is it less expensive to furnish 275 finned tubes, or 400 bare tubes? The deciding factor may not be tubes. A smaller bundle means that a smaller chiller housing can be used, reducing the cost of a chiller. A smaller housing means less stored refrigerant is needed to operate the system, lowering the total cost of the refrigerant and chiller. There is a downside in the use of finned tubes. If they are special in terms of production, availability, or material, they may be difficult to obtain in an emergency, while bare tubes are almost always available.

14.4 TURBULATORS

Turbulators are devices that are useful in improving the heat transfer rate of fluids flowing in the laminar or transition regions. They are almost always a tube side item. There is an exception. In double-pipe exchangers, longitudinal fins are sometimes cut at intervals and bent in such a way that part of each fin projects into a stream that is in laminar or transition flow, a feature that serves the same purpose as a turbulator. Their most common application is cooling liquids with high viscosities such as lubricating oils. To understand the purpose of turbulators requires an understanding of the "Tube Side Heat Transfer Curve," Figure 2-2.

When the Reynolds number of a flowing fluid is below 5000, the tube side rate in an exchanger can be improved as evidenced by the following example. Consider two different bare tube exchangers, one with 2-ft long tubes and the other with 30-ft tubes. The L/d factor in Figure 2-2 for a 2-ft, 1-in 18 BWG tube is $(2 \times 12)/0.902 = 26.6$. For a 30-ft-long 1-in 18 BWG tube this value is 400. Referring to the same curve and at a Reynolds number of 2000, the heat transfer factor j is 3.2 for 2-ft tubes and 7.8 for 30-ft tubes. In other words the heat transfer rate for the shorter length tube is (7.8/3.2) or 2.44 times greater than for the longer-length tube. A short-length-tube exchanger is usually not economical, while a longer-length unit is. The last two sentences have contradictory conclusions in terms of economics. The benefit of each can be attained using turbulators. The example is typical of conditions where turbulators can be useful.

Turbulators are designed to improve the heat transfer rate, allow the use of longer tubes, reduce an exchanger's overall cost, and make better use of pressure drop. In the example cited, the heat transfer rate is poor when the flow pattern is laminar for a relatively long distance as is the case using 30-ft long tubes. When the fluid turns at short intervals, as it does in the 2 ft tube example, the laminar flow pattern is broken in the turns. When the fluid enters another tube, it takes time and distance for laminar flow to again develop. The pattern is again broken after 2-ft of travel. The obvious need is to break the flow pattern using a method other than turns so that longer tubes can be used.

Many designs exist to accomplish this. Three are shown in Figure 14-1. Some are constructed of wire mesh; others of flat metallic ribbons with bends, crimps or twists added. The ideal is to break the laminar flow pattern

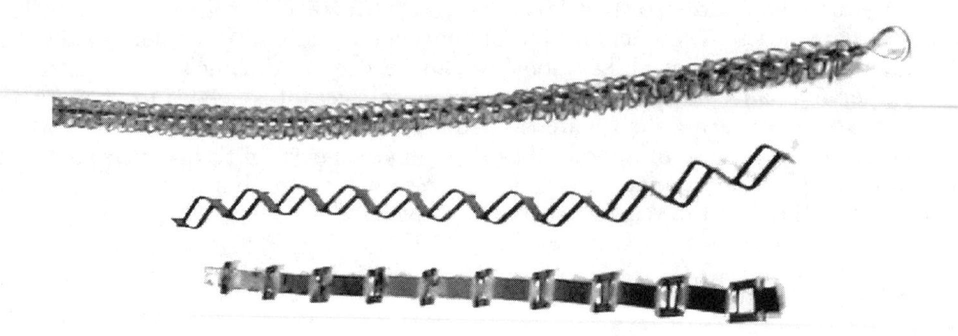

Figure 14-1. Various turbulator designs.

at the center of the tube where the velocity is greatest. Another need in designing turbulators is that they be held in position relative to the tube. A frequent way of doing this is to use an interference fit between tube and turbulator. The efficiency of turbulators improve when there is positive contact between them and bare surface, that is, when they are similar to fins joined to bare surface. In one case, flat bars twisted like ribbons were used. This design is not as efficient as others although there are applications where it is effective. This turbulator is difficult to hold in position and, when loose-fitting, have been known to rotate acting like a slow-speed drill. Some have bored holes in cover plates over time.

Consider the cut-and-twist fin commonly used in double pipe exchangers-a construction similar in concept to turbulators. Cutting a fin does little to improve the heat transfer rate. Its purpose is to allow the twist to be made. The twist places a portion of the fin in the laminar flow stream where it breaks the flow pattern creating the turbulence needed to improve the rate. This modification can reduce an exchanger's size by 50% or more depending on the specific application. A side benefit is that smaller exchangers often have shorter lengths of travel, hence, more pressure drop per foot of travel resulting in a further improvement in the heat transfer rate.

Referring to the example above, there is nothing limiting about breaking the laminar flow pattern at 2 ft intervals; turbulators often interrupt the flow pattern at $^{1}/_{4}$-in intervals or less which further improves the rate. See the designs in Figure 14-1 and the cut-and-twist of Fig 14-2.

Turbulators are not used on the shell side of a shell-and-tube exchanger. The nearly continuous expansion and contraction experienced by fluids crossing a bundle creates turbulence similar to that of turbulators. If the shell-side rate must be improved, consider using lowfins rather rather than bare-tubes. These have more surface to transfer heat and are easier to handle than turbulators. A similar comment can be made of plate exchangers. Fluids crossing herringbone-shaped plates are continuously subjected to turbulence, so there is no need to artificially create it.

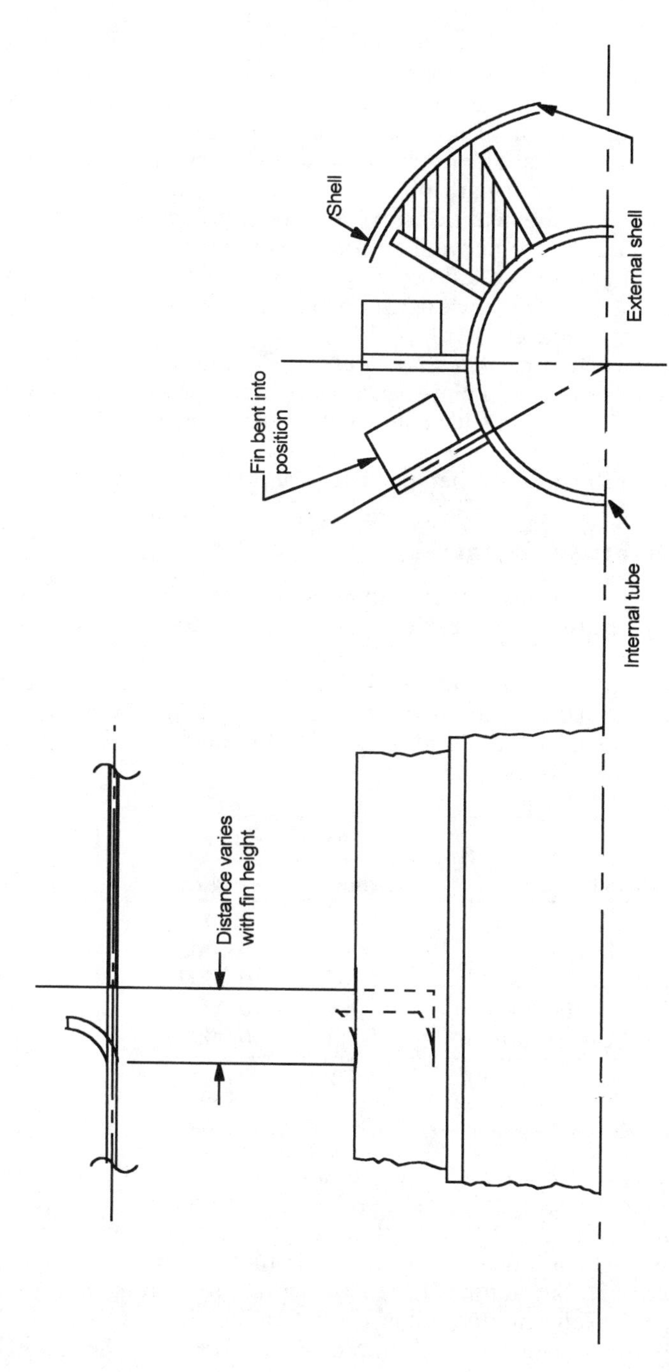

Figure 14-2. Longitudinal fin showing a cut-and-twist design. *Note:* Consider the shaded area between two flat fins. This area is similar to the inside of the tube flow area. A cut-and-twist added to the fin is similar to a turbulator added inside the tubes.

Turbulators are not recommended in fouling services because of problems associated with cleaning. Neither are they recommended in services where particles are present. They are usually not a spare part item. It sometimes happens that after their removal they are not reinstalled due to damage sustained during removal. They are usually too thin to correct damage done in the removal process. The omission of turbulators in the reassembly process will create a shortcut path through the tubes that lack turbulators resulting in a drop-off in the heat transfer rate. The problem is that the engineer will not be aware that some were omitted following cleaning.

In the interest of clarity, the following should be noted when comparing the use of turbulators with other designs. Particle size has created problems in exchanger designs (Sections 8.3, 15.2, 19.6, and 19.7). In general, turbulators cannot be used when particles are present. Cut-and-twist fins and the herringbone pattern of plates in plate exchangers break up laminar flow just as turbulators do. These designs have flow paths that allow particles to pass. The question of particle size may effect their selection but in most cases these designs will perform well when particles are present.

14.5 Automatic Brush Cleaners

To reduce fouling inside tubes, automatic brush cleaners can be used while the exchanger is operating. The brush systems are principally used in water or water/glycol services. Their advantages can be illustrated by an example. Similar resistances have been assumed for two streams, the only difference being that fouling is 0.001 without tube cleaners and 0.00025 with brush cleaners. This is typical of the cleaning value of brushes. Consider the following, with resistance values given in hr sq st °F/Btu:

	Brush cleaner	
Resistance	**Without**	**With**
Shell-side film	0.001	0.001
Shell-side fouling	0.0005	0.0005
Metal	0.0003	0.0003
Tube-side film	0.0009	0.0009
Tube-side fouling	0.0011	0.00028
Total R	0.0038	0.00298
U	263	335

By using brushes, the size of exchanger required is smaller by the ratio of these heat transfer rates, in this case, 21.5%. This ratio can be higher or lower depending on the effect brushes have on the tube-side fouling resistance. In a new installation, consideration should be given to buying a smaller exchanger with an automatic brush cleaning system to take advantage of the higher overall rate that follows.

Brushes may not be beneficial economically, as shown in the following example:

	Brush cleaner	
Resistance	**Without**	**With**
Shell-side film	0.035	0.035
Shell-side fouling	0.003	0.003
Metal	0.0004	0.0004
Tube-side film	0.0012	0.0012
Tube-side fouling	0.0011	0.00028
R	0.0407	0.03988
U	24.57	25.075

In this case the reduction in heat exchanger size using the automatic brush cleaner is 2%. The cost of a brush cleaning system probably cannot be justified.

A more common example is in improving the overall heat transfer rate of an existing installation. Suppose that after a unit has been in service it is found that its fouling rate is higher than estimated and, as a result, the exchanger is not meeting its performance guarantee. Some exchangers can be retrofitted with a brush system. The added brush system will keep tubes clean and help assure performance is met, but will not require modifying the existing exchanger.

In summary, three cases have been cited:

Case 1: A smaller exchanger can be selected to meet performance if an automatic brush cleaning system is furnished.

Case 2: In some cases the use of an automatic brush cleaning system cannot be justified.

Case 3: An exchanger's heat transfer rate can be improved, effectively increasing its capacity by using automatic brush cleaners without making changes to the existing unit. *Note:* If flow is increased, be sure the nozzles can handle the new flow.

Automatic brush cleaners operate as follows. A cage is installed at each end of the tube. In one cage a brush is installed. It acts in a propeller fashion as it passes through the tube. The brush is oversized, that is, there is interference between brush and tube. At around 8- to 12-hour intervals (each installation has its specific requirements), flow is reversed for a short period of time. When this occurs the brush's rotating action cleans the tube. A short while later flow is again reversed, with the brushes returning to their original position. This action is repeated at selected times in a given application.

Among the advantages of brush cleaners are that:

Maintenance costs are reduced.
Exchanger downtime is reduced.
Water treatment charges are often reduced, including the need for chemicals.
Water consumption is reduced.
The exchanger can be smaller in size.

14.6 SMALL-DIAMETER TUBES

One way of reducing the size or length of a shell-and-tube exchanger is to use small-diameter tubes, which allow more surface to be placed in a given volume (see Table 2-3). There are drawbacks to this approach. Small-diameter tubes are usually thinner and spaced on tight pitches, so that they reduce the flow that can be handled on the shell side. The chance of a flow blockage is increased if relatively large particles are present, as will be discussed in Section 15.2. An advantage of small-diameter tubes, particularly those with internal and external fins, is that when they are enclosed in the housing of another product such as a chiller, the chiller's housing can also be smaller, reducing cost.

14.7 ADVANTAGES OF PLATE EXCHANGERS OVER SHELL-AND-TUBE UNITS

Plate exchangers are built more compactly and are smaller in volume than shell-and-tube exchangers (see Chapter 8, "Plate Exchangers"). The surface is almost always made of a higher-grade material than the steel or nonferrous tubes of shell-and-tube units. Plate exchangers are competitive in price because, while higher-quality materials are furnished, less surface is required. (See Section 8.2.) Plates are thinner than are the tubes in competitive units. These exchangers are beneficial when the installation space is small and where less space is available for cleaning. They are not acceptable when operating pressures or temperatures are high. In roughly 90% of cases they are ideal where space is limited. The need for high fouling factor allowances tends to make them noncompetitive, but this requirement is relatively rare in most applications where plate exchangers can be used.

14.8 DESIGNS THAT USE FEWER FINNED TUBES PER LAYER AND MORE LAYERS

An air cooler is often selected using a narrow-width section and more tube layers. An alternative could be a wider section with fewer total tubes but more tubes per layer. If the former is chosen, the structure will be narrower, the diameter of the fans will be smaller, and the width of wind fence and louvers will be less, as will that of a recirculation chamber if it is needed. In total, the reduced cost of these items and of shipping will result in a lower-cost unit.

Almost the opposite is true in cases where the section is made wider (the exception is costs associated with wide shipping loads), provided the cost of energy is ignored. Fewer layers of tubes to cross reduce the air-side pressure loss, wider sections mean more air can cross, and more air means a larger

LMTD, with less surface needed. Said another way, the cost of the auxiliary equipment increases but the operating cost will be less.

Most equipment is chosen based on low initial cost and not low operating cost. If the latter is desired, be sure the cost of energy is given in the specifications. The narrower coil usually has the least initial cost. The wider unit usually has the lower operating cost, or said another way, is more energy-efficient. (See Section 17.5.)

Chapter 15

Techniques to Reduce Exchanger or System Cost

This chapter describes several ways to cut heat exchanger cost. To obtain these benefits, user and manufacturer must work together. After-the-fact conclusions are drawn following the examples in the chapter. Section 15.1 will illustrate how less refrigerant is needed to operate an A/C system if the evaporator and condenser use smaller-diameter tubes. Users should define the minimum acceptable tube diameter. In the case described in Section 15.2, a workable quench oil cooler could have been furnished had a flaking problem been recognized beforehand, while the success of the furfural cooler described in Section 15.3 could have been ensured by selecting equipment adaptable to frequent cleaning. These are the kinds of requirements that should be in the specifications, as is noting whether noncode units are acceptable. Defining or anticipating site conditions is the responsibility of the specification writer. Otherwise, the rating engineer has no way of knowing of these needs.

The other examples cited in this chapter are the responsibility of the heat exchanger manufacturer. To be competitive, manufacturers usually choose the lowest-cost tubing that meets the specifications. Users can influence a selection by advising the smallest diameter they will accept or specifying the size, gauge, and diameter they want. Shop assembly is almost always less expensive because it is costly to assemble and disassemble exchangers for shipment, to match-mark pieces, box parts, and keep track of items shipped. Manufacturers usually select factory-assembled units that can be delivered rapidly, have low energy usage (provided this is a requirement), use long tubes for economic reasons, and employ lined headers and channels if this construction is lowest cost. Before finalizing a selection, ask about requirements similar to these that may have been overlooked when preparing the specifications, as these influence the selection.

15.1 QUANTITY OF REFRIGERANT IN SYSTEM

The quantity of refrigerant in a closed system will be less if the volume in the exchanger(s) is smaller. Tube-side volume depends on the inside diameter of the tubes. The following example was prepared to illustrate this point. An evaporator with a given amount of bare tube surface has less volume as the tube ID becomes smaller (see Tables 2-1, 2-2, and 2-3).

Assume two evaporators each having 700 ft^2 of surface with the tube sizes shown in the tables. One uses $^3/_4$-in and the other $^1/_4$-in-OD tubes. The difference in the tube-side volume of these exchangers is 8.271 – 1.889 = 6.38 ft^3. The use of smaller tubes results in savings due to the reduced volume of refrigerant required, as follows:

Reduced refrigerant volume = 6.38 cubic ft^3
Approximate weight of refrigerant = (62.4)(1.08 sp. gr.) = 67.4 lb/ft^3
Approximate cost of refrigerant = $10/lb
Approximate savings in cost of refrigerant = (6.38)(67.4)($10) = $4300 (this is for the evaporator only)

This is an estimate, as refrigerants have different densities and costs per pound, but the lesson is that smaller-diameter tubes reduce the quantity of recirculating fluid required, thus reducing the cost of the system.

This reasoning also applies to A/C condensers. They are larger than the evaporator, so the savings are greater. For reducing system cost, smaller-diameter tubes should be considered unless there is a good reason for not using them. In this example the evaporator with the 1/4-in tubes might cost $1000 less than the one with 3/4-in tubes. Thus the savings would be $1000 plus $4300, or $5300 for the evaporator only. The savings on the condenser will be greater because it is larger.

Now consider a closed system that uses steam. Water is less expensive than refrigerants, and so no comparable savings result from using smaller-diameter tubes.

15.2 CASE STUDY: QUENCH OIL COOLER

A metal-stamping plant had identical shell-and-tube exchangers installed on several quench tanks. Their purpose was to remove heat acquired by the quench oil during the quenching process. The user wanted a second opinion as to why these units were not performing.

The quench oil was on the shell side, with water in the tubes. Metal pieces flaked off stampings following quenching. These were carried by the oil to the shell inlet, where they collected. The pieces were large enough to block the exchanger entrance, causing a flow reduction, enough to prevent coolers from performing.

The existing shells had 5/8-in-OD tubes spaced on 13/16-in triangular pitch. The spacing was too tight to allow metal flakes to pass. Both streams were high-fouling. The user advised that it took 2 men 8 hours each to clean one shell-and-tube unit. One unit per week was cleaned on weekends for 12 straight weeks. Cleaning was described as "the best that could be done." One weekend was idle before the cleaning cycle was repeated.

These units required an unusually high number of maintenance hours: 8 hours a day for 2 men on 48 Saturdays a year, or 768 man-hours per year minimum. In theory the units were sized properly, yet they did not perform in service.

The Standards of the Tubular Exchanger Manufacturers Association define nozzle impingement protection requirements at the bundle entrance.

The exchangers were not built to this standard, as it was not listed as a requirement in the specifications. No exchanger will perform if the inlet area does not allow full flow to the exchanger. The design (no dome area was provided) was inappropriate, because tubes partially blocked flow into the bundle. If the shell fluid is fouling and requires mechanical cleaning, the exchanger must be capable of being disassembled. Cleaning lanes should have been provided. Here, square pitch offers an advantage in that it provides paths for a steam lance to enter the bundle for cleaning purposes; triangular pitch does not offer this feature. Disassembling shell-and-tube units is time-consuming.

The challenge was to select an exchanger that performs while reducing maintenance time. Toward this end the following measures were proposed and accepted:

1. Cool with air rather than water and eliminate one fouling stream.
2. Place the quench oil — the remaining fouling stream — in the tubes for two reasons. First, the 3/16 in spacing that prevented flakes from passing is replaced with 1-in-diameter-tubes having large flow area passages thus minimizing the chance of blockage. Cleaning will be simplified using shoulder plugs because these are easily removed — less work than disassembling a shell-and-tube unit. The inside of tube surface is easier to clean than outside surface thus taking advantage of this known condition.

Using air for cooling had many beneficial results. Maintenance time was reduced from 768 hours per year to around 50. One of two air cooler sections remained onstream during periods of low production or when cool air was available (nights) to provide cooling while the other could be cleaned. The former were overtime hours; the latter were not. Another result was that cleaning the water side was no longer necessary. Then, during the winter, when fans were off, cold recirculating air provided free cooling. As an added benefit, one air cooler with two sections replaced several shell-and-tube units. The performance guarantee was met.

An improperly chosen exchanger design may result in reduced performance, difficult maintenance, or both as happened in this example. After the units were in service it became known that the quench oil cooler was not meeting performance. In switching to air cooling one fouling problem is eliminated. Three types of air cooler closures are available, manifold, cover plate, and shoulder plug. For this example the manifold header would not be acceptable because it does not provide access to clean the tubes. The question becomes, Which of the other designs, cover plate or welded box, should be used? In many cases, either is acceptable. Here are variables to consider.

Welded box headers can be used at higher pressures than cover plate headers because their welded-in pass partition plates can act as plate stiffeners. Tubes can be cleaned because removable shoulder plugs have been provided. If the header box must also be cleaned of grime and flakes, this could be achieved by adding clean-out nozzles (these must be listed in the specifications) on the welded box header. These are not effective on all welded header designs because they may not provide access to all passes. When reaching through a clean-out nozzle, one pass partition plate may block access to another. This design can be beneficial at high operating pressures. A better way, provided operating pressures are low (usually under about 100 psig), is to use the cover plate design. This construction is best at low pressure particularly for units with many tube layers. In this case, less time is needed to remove cover plate bolts than shoulder plugs. Before finalizing a cover-plate selection, make sure that the gasket design is simple. On many installations it is difficult to maintain seals because of complex gasketing. Another factor is that heavy covers make handling and maintenance difficult.

A word should be said about vertical pass partitions. These can be part of most cover plate designs though it makes their gaskets more difficult to manufacture. In the welded box design of wide sections, say 10 ft or more, it is difficult to make the seal weld at this partition because one does not have a good view of the weld to be made in a dark area, roughly 5 ft away, with limited access. A better choice may be to make another selection to simplify header construction. It should be noted that some manufacturers have developed welding techniques that solve this problem, but not all.

Conclusion: The ratings on the stamping plant's shell-and-tube quench oil exchangers were correct. However, the plant's engineers failed to inform the exchanger manufacturer that large flaky pieces were traveling in the oil stream being cooled. The resulting unintended obstruction of flow prevented the exchanger from meeting performance. (Chapter 19 describes a similar result where the air flowing to a screen room was reduced following a change to a more efficient filter.) Many selections could have been made for this application, but most would not have performed because flakes would block paths of small cross-sectional area. Hence, the presence of flakes should have been in the specifications. The units delivered required excessive maintenance. Their tubes should have been arranged on square or rotated square pitch and on a wider spacing for cleaning purposes. An inlet dome should have been furnished.

A logical question follows. Would competitive units solve the problem? Plate exchangers would not have been a good choice because the entrance to heat transfer surface is small and would be blocked by flakes. Double-pipe units could perform provided their fins were widely spaced and large flow areas were provided. Exchangers with lowfin tubes on tight spacing would block flow. On the other hand, nearly any shell-and-tube, double-pipe, or plate exchanger recommendation would have been different had the flaking condition been included.

15.3 AIR-COOLED FURFURAL COOLER SELECTION

Selecting a furfural cooler is much like selecting a quench oil cooler. Both fluids being cooled are viscous, both require that tubes be mechanically cleaned, both use the same sizing method, and both could be forced or induced draft. The difference is that furfural coolers are cleaned more often. It is unlikely the rating engineer would know this unless advised in the specifications. As a practical matter the governing condition is simplifying the cleaning effort.

Access must be provided for cleaning both tubes and header. This differs from the quench oil cooler, which did not require header cleaning. The cover-plate design was used for the reasons given in Section 15.2. This is the preferred construction when both operating and design pressures are low. The cover should have minimum weight for ease of handling. Designs with fewer tube layers have a lesser unsupported length of header dimension (Table 4-2). The smaller the cover width, the less will be the cover's weight. The less the unsupported length of the header, the higher the pressure it can sustain. Also, lower design pressures result in thinner covers, further reducing weight. Here is a practical action to take. Do not offer a manufacturer's standard low-pressure unit for this application. Rather, design for the operating pressure needed and reduce the cover thickness and weight accordingly. The maintenance crew will appreciate the reduced weight.

In this application the cleaning requirements and frequency should have been in the specifications.

15.4 CODE AND NONCODE UNITS

Section VIII of the ASME code covers the scope of the code and the vessels to which it applies. This standard contains mandatory requirements, some prohibitions, and non-mandatory guidance in the design of pressure vessels. Mandatory requirements must be met; good engineering judgment is essential as not every possible service is covered. Do not overlook the fact that local, state, and federal laws may also apply to the site at which the unit will be installed.

Sometimes it is difficult to determine whether code construction is mandatory or noncode construction will suffice. Noncode is usually less expensive and for this reason is often chosen provided it is acceptable for the application.

15.4.1 Noncode Units

There are industrial applications, many in repetitive service, that do not require an ASME code stamp. Generally they are integral parts or components of mechanical systems such as compressors, turbines, generators, engines, pumps, and pneumatic or hydraulic cylinders. The fluids cooled are, for the most part, water, water/glycol solutions, lubricating oils, and refrigerants. As a general rule in these applications large volume units require ASME code construction and are discussed in Section 15.4.2. Units not built to the ASME code are usually those having a small volume. There are exceptions as the volume of a vessel is not a governing code condition in most cases.

There are a number of reasons why smaller exchangers may not need to be built to code. In small exchangers, a leak to ground is generally not a hazard because the quantity that leaks is small in part because the fluids are nearly always in closed systems. The exceptions are refrigerants that contain chlorine such as R 11, R 12, R 504 and others. These vaporize at atmospheric pressure and migrate to the upper atmosphere. Free chlorine ions in these products are known to deplete the ozone layer. Leak prevention and a clean-up procedure are required. Minimizing a fire hazard may require that a unit be built to code.

Most noncode services use small-diameter tubes, that is, less than 5/8 in, but not always. They are often spaced on a tighter pitch than are those of Code units. For the same metal thickness, small diameter tubes can sustain higher pressures than large diameter tubes. Thinner tubes can be offered at lower cost, an economic benefit. Flat tubes can be used in services such as lube oil coolers and as radiators. Not only are their tubes thin but they minimize obstruction to air flow. To illustrate, refer to the example in Table 4-4 where the tube obstruction is 12 sq in and the fin obstruction is 1.98 sq in. If the 1-in tube were flattened to half this diameter, its flow obstruction would be reduced by half to 6 sq in allowing a greater quantity of air to pass. The need for code construction may still be necessary to avoid the possibility of fire or a change in tube material may alter the thermal expansion of some components relative to others to cite two examples.

Manufacturers of ASME code exchangers should not be surprised if their units are not competitive with commercial units made of thinner materials. Low cost noncode construction is adequate for many applications and usually will have the shorter delivery time. Manufacturers usually build either Code or noncode units but not both. Some offer either construction. Users should contact suppliers that build units to the construction they require. Commercial units are frequently used in repetitive services with the only change being the quantity of cooling required. Examples include air and water, jacket water and water, lube oil and air, lube oil and water, refrigerants and air, etc. For these applications, manufacturers frequently develop rating curves so that even unskilled personnel can select units from their standard lines.

15.4.2 ASME Code Units

The same repetitive applications may often require code construction, as do some others. As units become larger, code construction becomes more common. Not all repetitive applications are noncode. Code construction is likely to be needed on air or hydrogen coolers for generators; compressor intercoolers, whether air- or water-cooled; and ammonia refrigerant evaporators or condensers, air- or water-cooled. Depending on size of unit, code construction may also be necessary on lube oil coolers, other refrigerant evaporators or condensers, and lube oil coolers, in all cases either air- or water-cooled.

The following are reasons for selecting code construction. For a more thorough discussion of this subject, see the rules of the ASME code, Sections III and VIII.

1. Code-constructed units are usually built of high quality materials and seals. Thus they cost more than non-code units. Consider two types of services and their shutdown needs. Power plants often run up to 5 years without shutting down; refineries, on the other hand, often run 1 to 2 years before maintenance is needed. Hence, these plants almost always require code construction due to the consequential losses that result from an unwanted shutdown, e.g., financial loss, production loss, the possibility of fire, and personnel safety. Any condition that shortens these times is apt to be costly. In other applications the decision of construction to specify often depends on minimizing similar losses.
2. A combination of high design pressure and temperature (over 300 psig and 400°F) almost always requires code construction.
3. Temperature limits are difficult to define as they usually depend on the materials of construction. For example, double pipe exchangers can be used to a high temperature (650°F) without requiring a code stamp. They can be code stamped for a minimum extra. Plate exchangers cannot be used above 375°F due to gasket material limits. Shell-and-tube exchangers (packed head, shell side) are limited to 375°F for the same reason. Other shell-and-tube units can be used at higher temperatures depending on their design and materials of construction. These are reasons for selecting code construction. The construction to specify is usually made by the buyer or user's engineer or their consultant.

15.5 TUBING

The potential for reducing cost in tube selection is large for shell-and-tube units, air coolers, and double-pipe exchangers. Tube size and material often have a major effect on sizing equipment, and they *always* have a large effect on cost. For example, the following choices apply to 1-in 12 BWG steel tubes, though similar options may exist for tubes of any material:

1. Average wall or minimum wall
2. Seamless or welded
3. Hot- or cold-rolled
4. Domestic or foreign manufacture
5. Selecting an equivalent tube gauge, for example, a 1-in 11 BWG average wall for a 1-in 12 BWG minimum wall; this is generally an acceptable substitute based on cost and material thickness
6. Source of tubing—mill or warehouse

These choices affect cost. In theory there could be over a hundred combinations of choices, but practically there will be fewer. Small quantities can usually be obtained from a warehouse at a lower price. Selecting an economical tube reduces the cost of an exchanger.

The user or their engineer(s) usually selects the metallurgy of the tubes. They are familiar with the product and the metallurgy needed to handle it. Any change in the tube metal selection usually requires their approval.

There are many reasons for selecting a tube material other than steel. Some metals minimize corrosion while others reduce scaling at high temperature. Yet another metal will avoid contaminating a product with iron. Some metals provide tubing strength at high temperature. Whatever the reason, a common goal of manufacturer and user is to seek the lowest cost exchanger that will perform successfully.

One way of reducing cost is to substitute one metal for another. This requires user approval. Reasoning, similar to the following, applies. Consider using titanium in place of admiralty metal tubes. Titanium costs more than admiralty metal on a per pound basis. However, titanium's density is half that of admiralty metal at 300°F and its strength is near double that of admiralty metal at this temperature. Thus the weight of titanium tubes will be about one-fourth that of admiralty metal because these can be thinner in service. This makes titanium tubes competitive though the quantity purchase of materials has a large effect on price. Another consideration is that the lighter metal, titanium, is often preferred on applications such as mobile vehicles. Minimum overall weight is an advantage on highway trucks and airplanes for obvious reasons. In fact, users nearly always will pay extra for this feature. Similar reasoning applies to other tube materials.

Another option might be to offer a tube of higher quality metallurgy on some items. Here is an example of where this may be feasible. In a refinery, item (1) may require 4000 pounds of 4-6% chrome-1/2% molybdenum steel tubes, item (2) may require 2000 pounds of 2 1/4% chrome-1/2% molybdenum steel tubes, and item (3) may require 1500 pounds of 1 1/4% chrome-1/2% molybdenum steel tubes. Often, one can combine these quantities and offer the highest quality metallurgy for all three items because the increase in the quantity of the higher quality metal allows it to be purchased at a reduced price, less than if the three metals needed were purchased separately. Often this advantage can be refined by selecting metals that are generally available from warehouses because this may further lower the purchase price of these materials.

It might be well to compare the thin higher-grade tubing of the previous example with the plates in plate exchangers. In both cases, and for the same reasons, thin high-quality materials, whether formed as tubes or plates, make both designs cost effective, and both perform well in service.

15.6 SHOP ASSEMBLY

There are times when air cooler bays are too large to ship as assembled units. The question to be answered is, Should many small, fully assembled units in parallel be selected, or fewer and wider sections having structures and drives packaged for field assembly? A choice must usually be made on the basis of availability of skilled field personnel and economics. For example, suppose an air cooler is 48 ft wide and has 32-ft tubes. It could be built using three bays, each 16 ft wide. Few roads or rails allow 16-ft-wide structures to be shipped. Hence, this unit would have to be shipped knocked down for field assembly. A better choice might be to select four 12-ft-wide bays, which usually can be shipped assembled.

Other matters need to be considered. The larger the unit, the less the cost on a dollars per square foot basis. If a unit is to be field-assembled, it is usually factory-assembled first, run in, and knocked down with pieces match-marked and packaged for shipment. Losing a box and tracing its shipment is not uncommon. The use of a large shipping container should be considered as all parts are placed in one box minimizing the chances that they will be lost.

Smaller, factory-assembled units usually cost more on a dollars per square foot basis, but these minimize field assembly. Match-marking parts and boxing for shipment is eliminated. Knocking down and reassembling units will not be necessary.

Two innovative ways of simplifying this problem follow. For induced-draft units, assemble half a plenum on each of two sections. Arrange the sections side by side on a rack. The plenum pieces can be bolted together to minimize the field work needed. For forced-draft units, assemble the section, drives, and structures together. The structure columns can be shortened to stub columns; that is, each column is shipped in two pieces. In this way the height of the unit is reduced for shipping purposes only, and field assembly is minimal. This problem is relatively rare for shell-and-tube units, double-pipe units, or plate exchangers. When these become too large, smaller units can be selected and installed in series or parallel or a combination. Sometimes barge shipment can be used to advantage. After the choice of field or shop assembly has been made, the cost of shipping, handling, and lift capacity at the site may result in another iteration before a recommendation is finalized.

Conclusion: To simplify field installations, include data about the availability and skill of field personnel plus lift limitations.

15.7 RAPID DELIVERY

Exchangers or their parts are sometimes needed rapidly to get a plant back on stream. Such a need, when it exists, should rule out any selection whose parts cannot be available for several weeks. All exchangers do not have the same delivery times. Replacement bundles of readily available materials can usually be assembled in two weeks or less. Except for small exchangers, spare bundles are rarely stocked.

Double-pipe exchangers and their parts are frequently stocked. When a unit is needed immediately, it can be taken from an order in production to satisfy the emergency. Another unit could be placed in production a day or two later. In emergencies, however, this scenario is not always reliable.

An advantage of double-pipe units is that spare tubes can be kept on hand for immediate use, as can gaskets. If shutting down for cleaning is necessary, the possibility of installing a spare unit should be considered. An added unit in parallel could be cleaned while the others are operating.

If an exchanger is operating in an area known to be troublesome, spare parts should be on hand. Manufacturers have difficulty locating the special materials needed for parts during short turnaround periods. Air coolers might benefit from having a few spare tubes on hand. Spare gaskets are useful in shell-and-tube and plate-exchanger applications.

In conclusion, thought should be given to minimizing spares and delivery problems before they become an emergency.

15.8 ENERGY

The subjects of energy use and exchanger geometry are almost always worked as one problem because each has a large effect on the other. These subjects are discussed in Sections 14.8 and Section 17.5. One centers on how to reduce cost and the other on reducing energy needs.

15.9 MATERIAL AND CONSTRUCTION CONSIDERATIONS

One way of reducing cost is to minimize the use of expensive metals. Here are ways this can be done.

In shell-and-tube applications one of two flowing streams may require parts of higher-quality material. If possible, place this stream on the tube side, so that fewer parts of costlier materials will be needed. If the high-pressure stream is in the tubes, it may be possible to use thinner tubes. This commonly occurs in high pressure steam power plants.

Another technique is to design the exchanger using a carbon steel shell, tubesheets, and nozzles and line them on the side requiring the higher-quality material (usually the tube side). This lining will usually be 1/8 in to 3/8 in thick. For ease of construction it is preferred that base and lining materials be capable of being welded together. The metallurgy of the tubing is a separate issue (see Section 15.5).

Another way of reducing the cost of an exchanger is by confining the expensive material to a short length of shell and choosing carbon steel for most of the shell length. The procedure is shown in Figure 15-1, where the temperatures of both streams are given. The section of shell located at "X" is at a temperature between the hot and cold streams and must be of a material that can withstand the high tubesheet temperature that occurs on this short length of shell. The length of the section of shell required can be determined by considering this length as a fin and determining its tip temperature. It should be long enough that its tip temperature (where the connecting weld will be) is below the temperature limit of the carbon steel shell. The shell stub length should be increased about 1 in so that the heat of welding does not affect the strength of the joint at this location.

Sometimes exchangers in series require materials of high tensile strength at high temperatures. The cost of material rises as the operating temperature increases. The exchangers at intermediate temperatures can use materials whose metallurgy is between the extremes of the highest and lowest

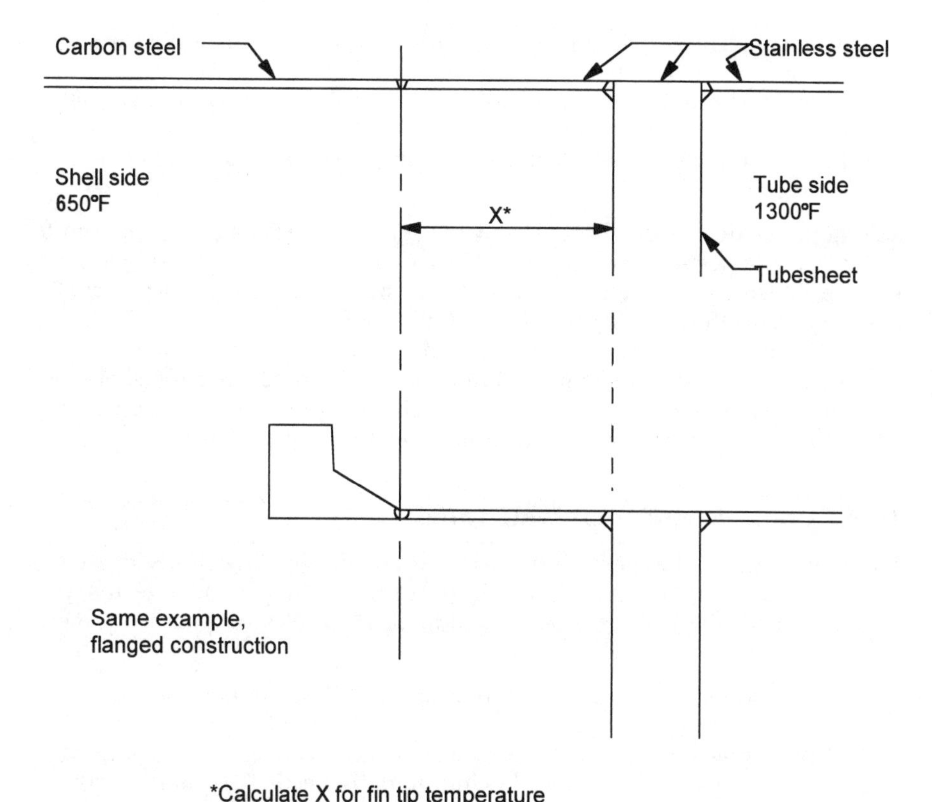

Figure 15-1. Minimizing the cost of an exchanger shell by reducing the use of expensive materials. (Welded and flanged connections shown.)

temperatures. For example, the tubes of a shell-and-tube exchanger may be 4 to 6% Cr-Mo steel above 800°F, $2^1/_4$% Cr–$^1/_2$% Mo steel above 650°F, and plain steel below 650°F. Under these conditions determining the lowest cost of a bank of exchangers is a trial-and-error process, for several reasons:

- The more units in series, the higher will be the LMTD correction factor and the lower will be the total surface needed.
- As the number of shells in series increases, the diameter of the shells required decreases.
- The larger the shell diameter, the less the cost of surface on a dollars per square foot basis, provided that the same tube diameter, gauge, pitch length, and material are used.
- If there are three shells in series, the higher-quality material(s) may be required in only one or two of the three shells.

- As an alternative, if four shells in series are chosen, only two may require the higher-quality metals.
- Another alternative, five shells in series, may require the higher-quality metals in only two of five shells.
- Be sure to include the selection of tubing, Section 15.5, which can have a large effect on price.

Selling prices developed through considering these factors are compared to arrive at the lowest-cost selection. The price may be further honed by replacing the intermediate shell with a higher-cost material whose price can be reduced due to the quantity purchase of materials.

Double-pipe exchangers face the same problem of material choice in high-temperature applications as described in the preceding for shell-and-tube units. These have the advantage of being much easier to price because they have far fewer parts and these are, in the main, standard.

15.9.1 Coolers with Dual-Wall Tubes

There are heat exchanger applications where the flowing fluids must not contact one another under any circumstances. In practical terms this means that a tube leak cannot be tolerated. Examples of where this need applies follow.

1. A fluid must be contained so that any leak that occurs will not cause injury to personnel or damage to the environment.
2. A fluid must be contained so that leaks cannot damage other equipment.
3. Positive separation of the flowing fluids is needed so that no violent chemical reaction can take place should a leak occur.

These requirement needs can be met using dual-wall tubes. Tubes can be of the same materials or of different metallurgies. Here is the way dual-wall tubes perform; refer to Figure 15-2.

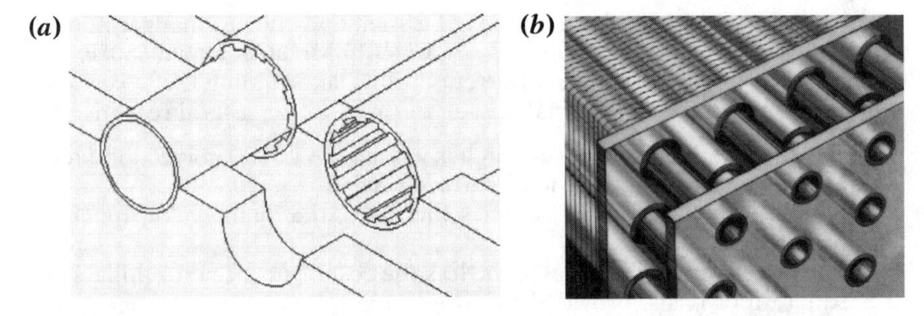

Figure 15-2. (*a*) Detail of double-tube construction typical of an air or hydrogen cooler for a motor or generator. (*b*) Cutaway showing double-tube bank. (Used by permission of BDT Engineering, Industrial Products Division.)

The external tube has internal serrations the length of the tube. The inner tube, positioned inside the outer tube, is expanded so that contact occurs over its full length. This contact allows the tube metals to transfer heat by conduction across their boundaries. The spaces between the serrations are flow paths so that, should a leak occur in either tube, it will be directed along these paths. An obvious advantage is that it allows for easy leak detection. Another is that an emergency shutdown can sometimes be delayed, allowing time to schedule repairs.

Here are applications where this construction is used. As an example of point 1, consider an exchanger having one stream of pure water and the other of radioactive water. A double-tube separates these streams preventing them from mixing should a leak occur. The next question is, How do we know if the inner or outer tube is leaking? If the fluid in the inner passage is radioactive, it is known that the leak is in the radioactive stream. Conversely, if the leak is pure water, the leak will be in the tube that contacts pure water. In this case, the emergency is not nearly as serious as is a leak on the radioactive stream.

A second service in which dual-wall tubes are used, point 2, involves salt water. The larger tube has fins on its external surface. Air or hydrogen gas, cooled crossing the fins, cools motor or generator windings. The goal is to keep salt water from contacting generator or motor windings. Dual-wall tubes are ideal in this service. If the pressure in the air or hydrogen gas streams drops and the passages are dry, it will be known that the leak is in the gas stream. Should the water side leak, salt water will appear in the inner passageways and be detected. This construction assures that salt water is kept from the windings.

15.9.2 Dual-Wall Plate Exchangers

Plate exchangers of double-wall construction have features comparable to those cited for dual-wall tubes as discussed in Sections 8.1 and 15.9.1. A typical plate is shown in Figure 8-4. For cooling motor and generator windings, the dual-tube construction is preferred because of the benefit fins offer in low-pressure gas applications. Dual-wall plates do not offer this feature.

Chapter 16

Adding Surface While Minimizing Downtime

There are times when an increase in exchanger surface is desired to upgrade a system's heat transfer capability. The design modifications that are described in this chapter provide benefits, where they can be used, because they allow a system to be kept on-line longer as surface is manufactured; the time during which units are off-line to make changes will be minimal. It should not be assumed that less work is involved. Rather, most of the work can be done while the system is operating. The changes suggested will not normally increase an exchanger's cooling or heating capability by more than 50% because of existing nozzle flow limits and pressure drop constraints. The exception is double-pipe units, whose capabilities can be increased by adding units in series or parallel.

16.1 ADD MORE PLATES IN PLATE EXCHANGERS

Under ideal conditions (i.e., when units have been selected for present and future conditions), surface can be increased by adding plates. Plate exchangers may not run continuously (8760 hours/year) but they can and often do. Plates can be added when a system is down. No change other than adding plates is needed and little production time is lost. Nozzles should be sized for maximum flow.

16.2 CHANGE FIN MATERIAL

The heat transfer rate in double-pipe exchangers that cool liquids can effectively be increased by 20 up to as high as 50% by changing the fin material from steel to either copper or aluminum. This option is not available to manufacturers that weld fins to tube because non-ferrous metals cannot be welded to steel. Non-ferrous materials have better thermal conductivity properties than steel (Table 5-2). To take advantage of this feature, all that is required is that the steel tube-fin elements be replaced with steel tubes with copper or aluminum fins. Elements could be manufactured while the plant is on line. It usually takes less than an hour to switch elements so plant downtime will be minimal. This is not an acceptable solution if the exchanger will be cleaned using a caustic solution. Aluminum in contact with this fluid does not hold up well.

This option is not available in air cooler applications. Air coolers rarely have steel fins (Section 5.2). They commonly use aluminum fins because of cost advantages (Section 5.1). Switching to copper fins usually increases the cooling capability of an air cooler by only 2 to 4% (Table 5–1). In other words, the benefits that can be attained using non-ferrous high thermal conductivity fins are already part of the construction of these exchangers.

Another air cooler consideration is this. Propeller fans used in process work are selected using 100% of the pressure drop available. This is rarely true of A/C evaporators which must meet the 500 ft per min velocity limitation to avoid blowing condensate off the coil (Section 25.0). The idea of switching fin materials to improve the heat transfer rate is a good one. Improving the rate to increase the available cooling also requires that the amount of air crossing the coil be increased. In most air coolers this cannot be done because the existing drive was selected at its maximum air movement potential. By contrast, most liquid coolers do not use all the available pressure drop allowed. Adding one or more sections in parallel is the best way of increasing an air cooler's cooling capacity.

16.3 ADD A PRECOOLER IN THE MAKEUP AIR LINE

In an air conditioning system adding a precooler in the makeup air line removes part of the cooling load from the evaporator, effectively increasing the evaporator size by this amount, but at a great savings in downtime, materials, and difficulty. To install a precooler in the makeup air line requires only two duct transition pieces, one for connecting the coil to the entering air duct and the second for connecting it to the leaving duct. The transition pieces and coil can be manufactured, assembled, and tested before installation begins. All that is required is to remove a section of duct and replace it with the preassembled duct sections and cooler. This assembly can usually be installed in eight hours or less, keeping downtime to a minimum.

Consider, on the contrary, directly adding surface to an evaporator. This usually requires a new coil, and space for a larger coil is seldom available. Equipment such as air filters and heaters must be relocated to accommodate it, requiring duct changes. Not only is the coil more costly, but the needed changes increase downtime. It is not unreasonable for this to take two weeks. The decision is, Is the cost of increasing the size of the evaporator worth the cost of two weeks' downtime, or is the makeup air cooler modification with its eight hours of downtime preferable? Be aware that the amount of inlet air precooling when added to the evaporator cooling is the maximum cooling that can be achieved by the system. If the cooling required is greater than this, adding a precooler will not meet the need. In this case, a larger evaporator will be needed.

This benefit may introduce trouble. Outside air directly contacts the makeup air coil, and freezing can occur on the water or air sides (or both) of this coil. Freeze and drain problems are more difficult than at the evaporator.

Under normal conditions, that is, no air precooler, makeup air is warmed by the main recirculating air stream before it enters the evaporator thus minimizing the chance of freezing at the evaporator. At the air precooler, there is no recirculating air stream to warm the entering makeup air thus increasing the chance of freezing at the precooler. An electric air heater at the preheater can be used to prevent freezing. At shutdown, it must be "on" in cold weather or the unit should be drained. This is easily said but difficult to implement particularly on week-ends when only a skeleton crew is available. The precooler design should be considered when the quantity of makeup air is increased, a condition common in older buildings being upgraded to satisfy OSHA fresh air requirements or for other reasons. The precooler's principle advantage is that it is an inexpensive way of increasing the cooling capacity of a system and it can be installed with minimum downtime.

16.4 ADD DOUBLE-PIPE UNITS TO A BANK OF EXCHANGERS

An advantage of double-pipe exchangers, particularly when an expansion is considered, is that units can be added in series or parallel to increase capacity. If a 50% increase in surface is needed, increase the units in series from, say, 2 to 3. This may increase the pressure drop on the shell side by more than desired. One option is to use the next larger diameter shell, that is, use units with 4-in rather than 3-in shells. These units are usually designed to mate with one another.

If tube-side pressure drop is a problem, consider a change similar to the following. The pressure drop for 3 sections in parallel and 4 in series can be reduced by changing to 4 shells in parallel and 3 in series. Similar patterns apply to both shell- and tube-side choices. This may result in new tube connectors, depending on the tube-side arrangement selected. The shell-side manifold need not be replaced, although a piece may be removed or installed to mate with new units. These changes can minimize downtime. The suggestion does not work in all cases, but it does in many, resulting in the rapid return of a plant to service.

16.5 REPLACE BUNDLES WITH SMALLER-DIAMETER OR LOWFIN TUBES

Another way of reducing turnaround time while adding surface to shell-and-tube units is to replace tube bundles only. More surface can be added by reducing the tube diameter or by replacing bare tubes with lowfins. New ratings will be necessary. New bundle(s) can be built while the plant is running. There are advantages to this approach: existing connections need not be disturbed, the cost of a new shell is avoided, and the shell is not removed to make the change.

Small-diameter and lowfin tubes reduce the tube ID, thus increasing tube-side pressure drop. This means the number of tube-side passes decreases. Changes to tube-side pass partition plates may be needed. This is a way of increasing surface at minimum cost and downtime. Check that nozzles will carry the increased flow.

16.6 AIR COOLERS

Surface can be added to air coolers by adding bays. Make sure that footings and concrete bases will handle any increased load placed on them. This applies particularly to the outside columns of cooler banks.

Another method is to select a unit that requires, say, 11 sections but furnish 12. Eleven sections would be on-line at any time while the 12th was valved off for cleaning. (The disadvantage is that all sections will require inlet and outlet flanges and valves.) This design provides continuous cooling without shutting down. In essence this could be called a spare-section design.

16.7 INSULATION

Air-conditioning systems are designed to provide cooling to given areas of buildings. A way of increasing capacity is not to change the cooling system. Instead, add insulation to the walls and roof of the building. This will reduce the cooling load on the A/C system, making the savings available for other purposes.

Chapter 17

Features That Should Make Selections Unacceptable

The subjects in this chapter could be deemed *good design practice*. Most but not all process and power plant heat exchanger specifications contain requirements related to the topics covered. When requirements are not specified—a common occurrence in commercial units—manufacturers should make selections in general agreement with the limitations found in this chapter. Some causes of trouble include too high a water velocity (pitting), geometry (won't fit in available space), use of too much water (existing piping cannot carry the quantity needed), high air velocity (blowing condensate off coil), and stacking (assuring seals do not leak) to name a few. Even then, omission of requirements from specifications can lead to disputes. A typical disagreement between might be, The unit delivered has an 80 dB(A) noise level while the plant standard is 75 dB(A). How can the rater know this or any other limitation if it is not in the specifications?

This chapter is intended as a reference to avoid trouble, for buyers to make certain that they are purchasing what they want, for sales personnel to assist in avoiding trouble, and for plant engineers to supply as complete a set of requirements as possible. Most of the topics touched on can be and have been the cause of trouble.

17.1 FACE VELOCITY (COIL) AND AIR VELOCITY (DUCT)

Air flowing across a coil often has a limiting velocity given in feet per minute. The limits for A/C systems and process coolers follow. They are not necessarily the same.

17.1.1 A/C Systems

Consider A/C evaporators. These are designed for an air-side velocity limit of 500 ft/min based on the face area of the coil (not the airflow area through the coil). The limit is based on experience. Higher velocities blow condensate off the coil, wetting nearby ducting and causing corrosion and cleaning problems.

Another consideration is that evaporator blowers must provide the static pressure necessary to move air through a coil and duct system. Process units use propeller fans and do not have a duct system pressure loss to contend with.

Evaporator air is confined by ducting. Its velocity in low-pressure ducts probably should not exceed 1200 ft/min, or a noise problem is apt to exist.

Noise travels in both directions while air moves in one. When air velocity exceeds this limit, ducts may have to be lined with noise-deadening material to lower noise. This is not always easily done, as the space available for duct and turns is often minimal. Duct noise should be addressed before a duct design is finalized. Retrofitting may be difficult because of space constraints. While, strictly speaking, this is not a heat transfer problem, be assured the heat transfer engineer will inherit it. *Caution:* Adding sound material to an unlined duct increases the air velocity in the duct and can make the noise problem worse as the lining reduces the airflow area.

The simplest solution is to address velocity in the design stage. Condensers, in most cases, are located out of doors and handling condensate is not a requirement. There is no ductwork to contend with. The 500 ft/min face velocity limit does not apply. Most condensers have propeller fans. These often meet the requirements of air-cooled process coolers, as described in the following.

17.1.2 Process Coolers

The 500 ft/min air velocity limit does not apply to process coolers, because there is no condensate to blow off coils. However, there are limits, namely, noise and energy. In sizing air coolers it is good practice to limit the tip speed of large fans to 10,700 ft/min. Above this value high noise levels are encountered (usually above about 80 dBa). Noise levels should be checked before a recommendation is finalized. Fans generate high noise when placed in poorly designed fan rings with large clearances between fan and fan ring. When noise rules are followed, the need for noise barriers is minimized.

In process work the face velocity is sometimes limited to 1,000 ft-per-min measured at the area along a row of tubes through which air passes. In A/C evaporators, the velocity limit is 500 ft per min measured at the face area of the coil. In process coolers the flow area through coils with 1/2- or 5/8-in-high fins on 1-in-OD tubes is approximately half that of the face area. Hence, the velocity limit may be nearly the same for A/C or process coolers. In the example the A/C evaporator measurement point has double the flow area and half the velocity of that of a process cooler. The process cooler's measurement point in this case has about half the flow area and double the velocity limit of an A/C evaporator. There is this difference, however. The 1,000 ft-per-min figure is not limiting in most air coolers used in process, ECU, and condenser services. In practice this value is exceeded. The 500 ft-min figure is limiting in evaporator service because blowing condensate off the coil must be avoided.

17.2 WATER VELOCITY LIMITS

Before sizing a water-cooled exchanger, the material of the tubes and the corresponding maximum water velocity should be known. The limits are determined from experience and are set to prevent pitting, which frequently occurs when the velocity limit is exceeded. Water velocity will affect an exchanger selection. The limit should be defined for refinery, chemical plant, and power plant applications. In these plants water is usually taken from a lake, river, or stream and passed through screens and filters before entering the exchangers. Typical velocity limits are:

Material	Max velocity, ft/s
Carbon steel	6 to 8
Copper base	Up to 7
Aluminum brass	8
Copper-nickel	10 to 12
Stainless steel	15

Another condition applies to each of the preceding materials. The minimum velocity should be in the 2.5 to 3 ft/s range, because this tends to keep tubes clean and prevent silt buildup.

As a rule water velocity limits are not specified in HVAC and electronics cooling applications because substantially lower quantities are required. They often use treated city water, while plant water is often taken from a river or lake. HVAC and electronics cooling equipment do not always run continuously as do process and power plants. Operating time can be in the 50 to 60 hours per week range. There are months when the equipment is idle. Some applications use treated city water. The recommendation is to know the source and quality of the water supply, the expected running time of the unit, and the velocity in your selection to see if a potential problem exists. When in doubt, specify limits.

17.3 NOISE

17.3.1 Fans

Many factors influence fan noise. Fan tip speed, tip clearance, blade angle, and shape of blades are among the variables. Two, tip speed and tip clearance, have a more prominent effect on noise than the rest. Fans deliver more air when the tip speed or blade angle is increased. Either change increases fan horsepower and noise level. If, however, the fan speed is lowered (with the same blade angle retained), the fan will deliver less air but at a lower noise level. This decreases the air-side heat transfer coefficient and makes additional surface mandatory. The problem of reducing noise is a trade-off. Slowing a fan delivers less air and lowers fan hp, noise level, and air-side heat transfer rate, but the exchanger needs more surface to meet performance. Generally speaking, the higher the horsepower of the fan(s), the less surface is required, while a lower fan horsepower means more surface is needed.

In high noise applications fan noise levels can be reduced by minimizing the clearance between fan tip and fan ring. While this clearance, however, can be maintained in the shop, it may be difficult to duplicate at the site. The fan ring tolerance needed for low noise levels is tight. Tight tolerances are expensive to maintain. The handling of fan rings during shipment and erection has an effect on tolerance. Other means may be necessary to obtain the desired tip clearance. Contact fan manufacturers for designs that address this problem.

17.3.2 Ducts

Duct air noise can be a problem. Noise will be minimal if the duct air velocity is 1200 ft/min or less. When an existing system's airflow and cooling are increased, more air must pass through the duct(s), which may cause a noise problem. Before making changes to the system, a fair idea (not always reliable, as it depends on the percentage increase) of how the increased flow affects noise can be obtained by removing the air filter (and its resistance) from the line. With less system resistance more air flows. The new flow is a single-point indicator of how the noise level increased and is an indication of whether it is tolerable. The test is simple. If the result is unsatisfactory, changes will be needed to meet the noise limit.

17.4 STACKING UNITS

Sometimes three or more shell-and-tube exchangers in series are necessary to obtain a reasonable LMTD correction factor. The problem that follows occurs when there are also three or more streams in parallel. Consider exchanger tolerances:

There are variations in the distance between the faces of inlet and outlet flanges.

1. There are variations in the amount of compression applied to the different types of gaskets. For example, if a ring-type joint gasket is used, it is more difficult to maintain a tolerance following compression than if another gasket design were used.
2. The variations apply to both streams of shell-and-tube units.
3. Dimensional tolerances of supports or mounting brackets also affect the installation. When supports are located too close to nozzles, the effect of one on the other may make it difficult to compress gaskets.
4. The cumulative effect of tolerances is different for nozzles and mounting brackets.
5. Manifold nozzle face dimension tolerances add to dimensional variations.

For these reasons there have been difficulties maintaining flange seals. To avoid cumulative dimension problems it is a good idea not to stack more than three shell-and-tube units in series. This reduces shimming, simplifies alignment, and minimizes field assembly problems. Installations with more than four units in series and three in parallel often have problems with leaks, seals, and shimming. This is not a common occurrence in double-pipe units, because smaller-diameter shells are more flexible except when the flange faces have ring-type joints. Seals are difficult to maintain when there are several units in series each with ring type joints.

17.5 CONSERVING ENERGY

17.5.1 Cooling Air

The heat transfer engineer is sometimes faced with problems similar to the following. A fluid is to be cooled from 160°F to 120°F. Air is available at 95°F. The question is, How much air should be used for cooling? Consider these possibilities. In cases 1, 2, and 3 it is assumed that the air temperature will rise 25, 20, and 15°F respectively. In each case the quantity of air flowing will be different which affects the LMTD and cooler size.

	Case 1		Case 2		Case 3	
	160	120	160	120	160	120
	120	95	115	95	110	95
	40	25	45	25	50	25
LMTD	31.9		34		36	

Assume the LMTD correction factor is 1 and the U value will be the same for all. The following applies, where W is the quantity of air flowing in lb_m/h and (0.24) is the specific heat of air.

In all cases the cooling load is the same but the amount of air required varies with the air temperature change.

Case 1: Cooling load = $0.6W(0.24)(120 - 95)$
Case 2: Cooling load = $0.8W(0.24)(115 - 95)$
Case 3: Cooling load = $W(0.24)(110 - 95)$

The surface required for case 3 will be 89% (i.e., 31.9/36) of case 1, but the quantity of cooling air required is 66.6% more, or (120 – 95)/(110 – 95) times the requirement for case 1. As the LMTD increases, the surface required decreases in the same ratio. The surface required for case 2 will be 94.4% (i.e., 34/36) of case 3, but the quantity of air required for cooling is 33.3% more, or (115 – 95)/(110 – 95) times as much as for case 3.

As more surface is added, the quantity of air needed is less. It follows that less horsepower or energy is needed to move less air. The horsepower required can be reduced further by rearranging surface, in patterns similar to the following. One can use five layers rather than six, or four layers rather than five, etc. As an example, a 10-ft-wide six-layer section will have almost the same surface as a 12-ft-wide five-layer section. The energy required to move air across five layers will be less because the flow area is 20% larger and the length of travel is 20% less. This variation should be kept in mind when trying to make a low-energy use selection. If an exchanger requiring more air is chosen, determine if this causes problems with horsepower, noise, or starving fans. [See also Face Velocity (Coil) and Air Velocity (Duct), Section 17.1.]

17.5.2 Cooling Water

A similar problem occurs with water as the coolant. Again, consider three possibilities:

	Case 1		Case 2		Case 3	
	160	120	160	120	160	120
	105	80	100	80	90	80
	55	40	60	40	70	40
LMTD	47.1		49.3		53.6	

The difference in the LMTD between cases 2 and 3 is 8.7%. To gain this advantage, the quantity of water pumped is doubled: (100 – 80)/(90 – 80). In case 2 the water temperature rise is 20°F; in case 3 it is 10°F.

Horsepower can be reduced by pumping less water. Often one of the following is the governing factor:

- The existing piping has to be large enough to carry the increased flow.
- The lowest-surface exchanger leaves little flexibility for future requirements.
- Compare the water needed for cases 1 and 3. For 13% more surface, or (53.6)/(47.1) times the surface of case 1, the water needed will be only 40%, or (90 – 80)/(105 – 80), of that required to operate the smaller exchanger. In many cases a larger exchanger can be justified economically. This also applies to process streams except that in most of these the LMTD decreases rapidly and the required surface increases rapidly so the effect isn't as pronounced.

For a discussion of cost and energy, see Section 14.8. A low cost selection may be an unacceptable solution for the following reasons:

- Too noisy
- Flow limited by a pipe's capacity to carry water
- Starving fans
- Space and shipping limits
- Delivery needs
- Cost

17.6 DIMENSIONAL LIMITS

An exchanger selection may be excellent from a thermal point of view but unacceptable on account of dimensional limits. To satisfy all requirements it may be necessary to make several selections before a recommendation is finalized. The following, all discussed elsewhere in this book, illustrate some conditions that impose dimensional limits on a selection:

1. Space needed to support auxiliary equipment (Section 2.1)
2. Available space on a pipe rack (Section 2.1)
3. Available access space (Section 2.1)
4. Tube length economics (Section 2.4)
5. Width considerations (Sections 21.5, 21.7, and 21.8)
6. Aspect ratio (Section 6.1)
7. Manufacturing or weight limits (Chapter 13)
8. Increasing the capacity of a plant without major changes (Chapter 16)
9. Selecting smaller exchangers to stay within a plant's lift capability (Chapter 13; Sections 2.1 and 21.6)
10. Choosing wider tube spacing for cleaning purposes (Section 15.2)
11. Conserving energy (Section 17.5)
12. Cleaning (Section 2.2)
13. Fan coverage and bundle geometry (Section 6.1)
14. Tube diameter and refrigerant storage (Section 15.1)
15. Tube and baffle hole tolerances (Section 19.14)
16. Tight packaging in electronics (Section 30.3)
17. Screen room ducting (Section 19.3)
18. Rack spacing (Sections 28.5, 30.5, and 30.9)

17.7 LIQUID FLOW ACROSS FINS

Air and hydrogen gas are the principal fluids that flow across tubes having high aluminum or copper fins (1-in tubes with $^1/_2$- or $^5/_8$-in fins). Such fins are of relatively soft materials that would easily bend or warp were they to be subjected to high liquid flows. There would also be a problem of maintaining the bond between tube and fin. Extruded fins are sturdier than wrapped fins. Even so, neither construction is commonly used with liquid on the fin side. It happens that the high outside/inside surface ratios (up to 20 to 1) they offer is rarely an advantage in liquid-liquid service.

In liquid to liquid service, integral fins and tubes are preferred, because of the dependability of the tube-fin bond and the outside/inside surface ratio they offer, usually between 3 and 7 to 1. Another advantage is that dependable tube supports are rarely a problem. It does not follow that high fins are never used with liquids on the fin side; they are, particularly as tank coil heaters (spiral fin units), for the following reasons:

1. The fins are of a strong material, steel.
2. The flow rate across the fins is low.
3. Fluids heated in tanks are often viscous. Outside/inside surface ratios of between 8 and 20 to 1 are common. The heat transfer rate of steam is commonly 15 to 20 times that of heavy hydrocarbons, making this construction favorable.

Conclusion: High wrap-on or extruded aluminum fins are generally not recommended for cooling or heating liquids on the fin side. The choice of fin is influenced by the application, the available pressure drop, and whether the fluid in contact with the fins is a liquid or a gas.

Chapter 18

The Significance of Metal Temperature

This chapter contains examples demonstrating the importance of metal temperature. Not all apply to heat exchangers, but all are heat-related problems that can affect the operation of process, air-conditioning, or electronics cooling facilities. The metal temperature is calculated much the same way in all cases.

In heat exchangers the tube metal temperature will be between the hot- and cold-side fluid temperatures. Its reading at any point will depend on the film and fouling resistances of one stream relative to the comparable resistance(s) in the other stream. If one fluid is steam with a low film resistance and the other is asphalt with a high film resistance, the tube temperature will be nearer the steam temperature. On the other hand, if one stream is cold water, with a minimum film temperature, the metal temperature will be nearer the water temperature. Knowing the metal temperature is necessary in many applications. Here are some of them.

18.1 SIGNIFICANCE OF TUBE-WALL TEMPERATURE IN SIZING AN EXCHANGER

Knowing the tube-wall temperature is often necessary for sizing an exchanger. The following example illustrates where knowledge of this temperature is useful and how it is determined.

Example:

Assume a 45° API hydrocarbon oil is cooled in an exchanger from 400°F to 150°F. Water at 85°F is available for cooling and leaves the counterflow exchanger at 105°F. The oil properties are given in Table 18-1. The log mean temperature difference is:

	Hot terminal	Cold terminal
	400°F	150°F
	105°F	85°F
	295°F	65°F
LMTD	152	

Table 18-1 Properties of 45° API Oil

	Temperature, °F		
	400	**244**	**150**
Thermal conductivity k, Btu/(h · ft · °F)	0.075	0.079	0.0815
Specific heat C, Btu/(lb_m · °F)	0.66	0.575	0.52
Viscosity z, cP	0.22	0.45	0.88
ρ (sp. gr.)	0.66	0.73	0.77

For these conditions the tube wall temperatures under various operating conditions were calculated and are given in Table 18-2. Calculations were made for six different operating conditions to cite the benefits of knowing metal temperature and some problems related to it.

1. Column 1, Hot Terminal (clean) At the inlet it is assumed that both sides of the exchanger have clean surfaces. If the hot fluid should enter the exchanger without the cold water "on", the tube metal temperature will soon reach that of the entering oil, 400°F. This temperature plus a factor of safety, say 25°F, is used to establish the design stress of the tube. When the water is turned "on", the tube metal temperature

Table 18-2 Tube-Wall Temperature T_{tw} Variations in an Oil-Cooled Exchanger

Resistances r, R in units of h · ft² · °F/Btu

Heat transfer rate U in units of Btu/(h · ft² · °F)

		At caloric temperature				
Resistance	**Hot terminal (clean)**	**Both sides fouled**	**Tube side fouled, shell side clean**	**Shell side fouled, tube side clean**	**Both sides clean**	**Cold terminal (clean)**
r^f_s	0.00190	0.00270	0.00270	0.00270	0.00270	0.00370
r^d_s		0.00200		0.00200		
r_m*	0.00034	0.00034	0.00034	0.00034	0.00034	0.00034
r^f_t	0.00095	0.00100	0.00100	0.00100	0.00100	0.00100
r^d_t		0.00256	0.00256			
R	0.00319	0.00860	0.0066	0.00604	0.00404	0.00514
U	313	116	151	165.16	247	194.5
T_{tw}, °F	224	161	182	125.7	142.4	103.2

*Metal resistance for 1-in 12 BWG tubes, steel, average wall.
r^f_s resistance of oil film, shell side
r^d_s resistance of dirt (fouling), shell side
r^f_t resistance of film, tube side
r^d_t resistance of dirt (fouling), tube side
r_m resistance of metal

drops to 224°F. The water "off"start-up condition, raising the metal temperature to 400°F, has been the cause of U-tubes cracking due to thermal expansion particularly if they are made of copper. If the water is guaranteed to be "on" at all times, the tube metal temperature can be used to determine it's stress value. In this case, the metal temperature difference is 176°F (400°F–224°F). This metal temperature range can be considerably higher in some refinery applications where there may be 3, 4, or 5 exchangers in series operating over a typical temperature range of 800°F (1000°F–200°F). The tube metal temperature plus a factor of safety is used to select the tube material of each of the exchangers in series.

2. Column 6, Cold Terminal (clean) The cold start-up condition and is used to calculate the exchanger's tube outlet metal temperature in the same way as for the inlet temperature. If the hot fluid is turned "on" first, the outlet area will reach 400°F after a short period of time creating the same problem as at the inlet. This operating condition is rare. The condition that should be noted is that the tube metal temperature is 103°F when clean and 161°F when fouled (column 2). Exchangers are sized for fouled conditions. Flow through the exchanger could change from turbulent to transition to laminar before reaching the exchanger outlet. The tube metal temperature (in every exchanger in series) is used to determine the exchanger's size and if the flow pattern changed while in the exchanger.
3. Columns 3, 4, and 5 show variations in metal temperature, 116°F to 247°F, that can occur for various values of fouling. Do not assume these calculations are purely theoretical. Many exchangers have been designed using high fouling factors. When disassembled, their surface is clean with no fouling present because safe fouling factors were chosen.

In the example the average tube metal temperature varies from 103.2°F to 224°F depending on location within the exchanger and the existing conditions. It can, under still other conditions, vary more. Usually, the variation is greatest at the hot terminal. When the tube-side inlet and outlet temperatures (not metal temperatures) differ by more than 250°F in one exchanger, the design should be reviewed to determine if a thermal expansion problem is likely to exist. Problems occur mainly at the inlet header.

From the data it is observed that as fouling (and therefore fouling resistance) builds on either stream, the metal temperature changes, and the magnitude of the change depends on the magnitude of fouling. It can change for other reasons as well, including reduced flow on either stream, which also changes film resistances. Other changes might include colder water in winter, the use of a different tube material, or a reduced heat load Q. Good judgment is necessary when determining these variables so as to prevent the tube stress from exceeding its design limit.

Here are reasons why it is beneficial to know the tube-wall temperature:

1. At the hot terminal, the wall temperature is a maximum. This, provided the flow is "on" at all times, and a factor of safety establish the

design temperature. The design stress at this temperature is used to calculate tube thickness. Often, a more expensive material on a cost per pound basis is the least expensive because thinner but stronger tubes requiring less metal can be used.

2. At the caloric temperature, the wall viscosity has a major effect on the film resistance of the viscous stream. It can be 3 to 5 times higher for a fluid being cooled, or it could result in an exchanger's being $^1/_4$ to $^1/_3$ the size, when heating liquids, of what it would be without the μ/μ_w factor included. This happens in steam heating applications. The factor will be near 1 in many applications (see Section 18.4).

18.2 CONDENSER DRIVING FORCE

Assume a condenser uses water to condense steam. If the exchanger is sized correctly, the steam pressure loss through the exchanger is often said to be "nil." Reporting the pressure loss this way is generally but not always correct. Here's why.

Assume that in a condenser steam is at 350°F (134.63 psia) with a metal temperature of 250°F (29.825 psia). A large force exists to drive the vapor toward the tube (134.63 psia – 29.825 psig = 104.8 psi). If the gas laws were followed, the pressure loss might calculate as 0.1 psi. This is insignificant compared with the driving force available. Hence, listing "nil" is correct. It will not always be nil, however, as shown in the following example.

18.3 HEATING WITH DESUPERHEATED STEAM

Assume that in Section 18.2 the tube metal temperature in the desuperheating zone is 350°F, a typical power plant value. If the tube wall is dry (no condensation), the superheated steam stream is sized as a gas. The advantage of a desuperheating zone is that water in the tubes can be heated above the steam condensing temperature, making the system more thermally efficient. The differential pressure driving force described in the previous section does not exist in this case.

In some processes, particularly food, the heat in superheated steam is used to control temperature. It is known from experience that heat delivered in small rather than large amounts is easier to control. This is illustrated by the following comparison. When condensing steam is used to heat a product, the number of Btus added to the product will be approximately 950 Btu/lb_m, depending on pressure. If a pound of steam superheated 100°F is used for the same purpose, the heat added will be approximately

$$(1)(0.45)(100°\text{F of superheat}) = 45\ \text{Btu/lb}_\text{m}$$

This is roughly 5% of the heat of condensation for steam. The skill is to design an exchanger taking advantage of this differential so that the tube wall will be warm enough to prevent condensation. Otherwise, steam will condense and large quantities of heat will be transferred.

18.4 BULK VISCOSITY AND TUBE-WALL VISCOSITY

The most frequent reason for wanting to know tube-wall temperature is to size an exchanger. Bulk and tube-wall temperatures can be substantially different. In heating situations the μ/μ_w factor can reduce the size of the heater by 65 to 75%. When liquids are cooled this factor can increase the size of the exchanger required by 50% or more. Gases are a different story. When a gas is heated, the size of exchanger required is increased; when a gas is cooled, exchanger size is decreased. (For the effects of wall viscosity see Sections 7.2 and 18.1.)

18.5 CHANGING METALLURGY

Steel is the metal of choice for storing or cooling concentrated sulfuric acid, provided the acid is below 200°F. As hot sulfuric acid cools, its metal temperature should be known so that the exchanger's tube material can be changed from a higher-quality metal, usually stainless steel, to the less expensive material, steel. Often the quantity of expensive tubes needed can be minimized by using two or three smaller exchangers in series. This way only one shell, or possibly two, requires the higher-grade tubing. To make this decision the tube metal temperature must be known.

18.6 SUBCOOLING

Consider an air-cooled steam condenser having six layers of tubes. In the usual case, the first pass will have five rows for condensing. The remaining row is in the second pass and is used for subcooling. The metal temperature in this pass should be known. If its temperature drop is over 50°F, it should have its own floating header at the outlet (i.e., the exchanger needs to have a split header; see Figure 18-1) to account for thermal expansion and contraction. Do not overlook the fact that subcooling is greater in winter than in summer because colder air cools the condensate further. It also magnifies the difference in tube expansion between passes. For these reasons tube metal temperature should be known.

18.7 PRODUCT TEMPERATURE TOO HIGH

There are applications in both the petroleum and food industries when it is necessary to keep the wall temperature below a certain value. In tank heaters,

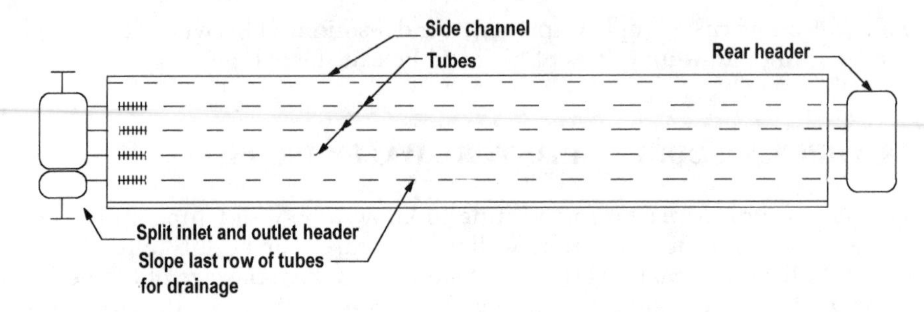

Figure 18-1. Air cooled heat exchanger with split header.

high wall temperatures break down heavy hydrocarbons and vaporize the lighter ones, causing flashing on the suction side of the pump. This causes pumping problems and product damage. The chance for this to occur can be minimized by keeping the tube metal temperature low.

Consider this example. Fuel oil is heated in a tank from 50°F to 120°F using 250°F steam. The expected resistances are in h · ft^2 · °F/Btu:

	Bare tube	Finned tube
Shell-side film resistance	0.045	0.045
Shell-side fouling resistance	0.003	0.003
Metal resistance, 3/4-in Sch. 40 pipe	0.00042	(NA)
Metal resistance, 3/4-in Sch. 40 pipe, sixteen 1/2-in steel fins	(NA)	0.020
Tube-side film resistance, steam (0.0005 corrected to outside surface)	0.00064	0.0041
Tube-side fouling resistance (0.0005 corrected to outside surface)	0.00064	0.0041
R	0.04970	0.0762

If the metal temperature of the product stream is calculated, the following values result:

$$\text{Bare tube: } T_w = 250 - \left[\frac{0.00042 + 0.00064 + 0.00064}{0.004970} + 102\text{°F LMTD}\right] = 246.5\text{°F}$$

$$\text{Finned tube: } T_w = 250 - \left[\frac{0.020 + 0.0041 + 0.0041}{0.0762} + 102\text{°F LMTD}\right] = 212.3\text{°F}$$

While the above values change with the application, and according to whether tubes are clean or fouled, the lesson is that when a product is heated, the tube-wall temperature will be less if extended-surface tubing is used.

18.8 REBOILERS

Reboiler sizing is beyond the scope of this text. However, it should be noted that when high temperature differences exist between the product and tube, a hot tube can cause a product to boil. As more heat is added, a vapor barrier is created between the tube and product. This film reduces the amount of heat that can be transferred, causing sporadic operation. In this and similar cases the tube metal temperature or the heat density must be known to correct the problem.

18.9 THERMAL EXPANSION IN U-TUBE EXCHANGERS

There is a difference in tube-wall temperature and the magnitude of thermal expansion in the legs of U-tubes in shell-and-tube exchangers as opposed to the legs of double-pipe exchangers. To illustrate this, assume the hot fluid is on the shell side of a shell-and-tube exchanger and enters the shell beyond the U-bends. After this fluid crosses the tube bundle one time, both legs of the U-tube experience roughly the same temperatures as the mixed temperature of the shell fluid contacts one side of the U-tube and then the other. This tends to balance tube metal temperature in both legs of the U-tube.

This is not true of a double-pipe exchanger in true counterflow. The colder portions of both streams are in one leg of the exchanger; the hotter portions are in the other. Hence, one leg of the U-tube will expand more than the other. This can be a problem at start-up, particularly if the hot stream is turned on before the cold stream. The leg carrying the hot fluids can expand considerably before the hot fluid reaches the colder leg. The magnitude of the problem is increased if the tube metal is copper, which expands 40 to 50% more than steel. Further, copper has only about 30% of the strength of steel at 300°F. A frequent result is that the tube(s) cracks at the U-bend. If frequent start-ups are necessary, consider using a tube of stronger material with a lesser coefficient of expansion than copper. Do not overlook the start-up metal temperature during the winter months, which can be well below 0°F, exaggerating the problem. This problem is more difficult if the inlet is located near the tube sheet because one end of the U-tube will be at a higher temperature than the other.

18.10 RADIANT ENERGY AFFECTING A ROOF OR CANOPY

On a cloudy day the ambient temperature on a roof or canopy may be at the design temperature, say, 95°F. On a sunny day, if the canopy is of light color, its surface temperature can rise 20°F due to radiant energy and another 15°F if it is of dark color. This raises nearby temperatures accordingly.

Security cameras at industrial sites may be under a canopy. The usual camera temperature limit is 104°F (40°C). Under metal coverings painted with a soft white or off-white color, midday radiation can raise the roof or canopy temperatures to near 130°F and the camera temperature to near 104°F. Changing the paint to a dark color may cause the camera temperature to rise further causing a failure. The temperature of the canopy is calculated the same way as tube-wall temperature.

18.11 TANK SHELL

Industrial fluids such as refrigerants, propylene and others in tanks or heat exchanger shells present a particular type of problem. Consider refrigerant 134a as typical of this grouping. If it is stored at 100°F, its pressure is 138.8 psia (124.1 psig). However, if the storage tank is subjected to the sun's radiant energy, its surface temperature could rise 35°F due to heat from this source. At this temperature, refrigerant 134a's storage pressure is 228.25 psia (213.55 psig). If 25 psig is added for safety purposes, the storage vessel should be designed for 238.55 psig (not 138.8 + 25, or 163.8 psig). This is another reason why metal temperature must be known. If the tank is buried, the lower design pressure will apply.

18.12 CABLES AND CIRCUIT BREAKERS

At temporary sites and similar applications, electrical cables are laid on the ground. The effect of the sun's radiant energy is a factor in the cable's ability to carry current. As the temperature increases, the cable's current-carrying ability drops. The cable's temperature is determined the same way as that of a heat exchanger tube. To carry the same current at elevated temperature, larger-diameter cables often must be furnished. Thermal insulation alone does not solve the problem. After lying on the ground for a time a cable's temperature is the same throughout regardless of the insulation used.

A circuit breaker will trip, regardless of cable size, if it reaches its current-limiting temperature when subjected to radiant heat. An example would be mounting a circuit breaker on a steel beam subject to radiant energy. To prevent the circuit breaker from reaching the beam's temperature, insulate the space between the circuit breaker and the beam. A more effective step is to mount the circuit breaker under a protective cover to avoid direct radiation from the sun. There are cases where both steps have had to be taken.

Chapter 19

Thermally Correct Systems That Do Not Perform

19.1 IDENTIFYING THE PROBLEM

It is not uncommon for heat transfer engineers to be asked why a system is not performing, even when the equipment was not sized by their organization. Minimum data may be all that's available. This chapter will provide the engineer with ideas about what to look for when units do not perform.

A related question is, How does one *know* when a unit is not performing? In refineries, chemical plants, and power plants, underperformance is known by measuring output. In air-conditioning systems it is known when rooms become hot and uncomfortable and users ask why. In electronic systems, problems become known when equipment fails and immediate attention is required. In efforts to get a plant back onstream quickly, it should be understood that the best solution is often overlooked in favor of the expedient one.

Industries differ in their approach to correcting failures. Power plants, refineries, and chemical plants cannot normally be altered without shutting down for safety reasons. Rack-mounted electronic systems (provided they can be isolated) and/or air-conditioning systems are or can be modified while the system is operating because there is little danger to personnel as changes are made. Very commonly, many changes will have been made but not documented; hence, site trips may be needed to determine why units are underperforming. Installation pictures can be beneficial.

Process exchangers are often fitted with couplings for pressure and temperature gauges. From readings taken at these fittings or from plant output it can be determined if an exchanger is meeting performance.

In selecting exchangers, consider factors in addition to flow, temperatures, and properties with the intent of avoiding future operating problems. Failures that have occurred in practice are cited in this chapter, and conditions that existed during operation are given. The intent is to instruct engineers and others about what to look for to avoid similar problems. In most cases the factors that caused the failures were not in the specifications or were overlooked; the neglected requirements will be identified. (In other cases a selection was not a good choice for other reasons.)

Reflect on the idea of who is responsible. Typically, two engineers, one representing the buyer and the other the seller, meet and agree on what is needed to satisfy a heating or cooling problem. A variation would be a seller's engineer meeting with a buyer's representative who is not an engineer. In either case one might assume the two meet as equals in agreeing on the equipment required. This may not be true. Consider the seller's heat transfer engineer, who may represent but one product, say, shell-and-tube heat exchangers. The equipment he proposes is the best his company can offer. Another product, not manufactured by his company, may be better suited for the application. The buyer's representative probably would not learn this speaking with the shell-and-tube exchanger engineer. Similar conditions apply to other products.

The buyer's engineer is expected to know the choices available. A buyer who works for a process, chemical, or power company should have knowledge of such products as shell-and-tube units, air-cooled spiral-fin units, air-cooled plate-fin units, double-pipe exchangers, spiral exchangers, plate exchangers, and others. Also, as examples in this chapter will show, other equipment in the plant often affects the performance of the chosen exchanger. It is necessary to understand many factors, and to be knowledgeable of auxiliary products, affecting exchangers such as structure, ducting, filters, heaters, noise, damping devices, manifolds, supports, fans, blowers, motors, V-belt drives, and gear reducers, to name but a few. The buyer's engineer is responsible for developing the requirements and specifications defining the equipment required.

Before beginning the task of correcting a problem, observe the following rules, which are nearly always beneficial:

- Never assume a manufacturer's proposed unit and performance guarantee are correct, although they probably are. Prorate the choice against units known to perform. In this way, an undersized (or oversized) unit can be identified early as the cause of a problem.
- Confirm that the equipment is connected as intended (see the discussion of tank heaters in Section 19.15).
- Review the manifolding, as this is an easily identified cause of problems.

Before proceeding through the rest of the chapter, you may prepare for the examples by thinking a certain way. When reading about a failure, suppose it had been *your* project. First, ask, "How could I have prevented this problem from occurring?" Then ask, "When the problem was identified, what should have been done to correct it?" Two lessons can be learned. One is that site conditions are often overlooked when specifications are written. Significant facts or conditions or those thought to be unimportant become fully understood only later. Even then, the information is often not passed on to the engineer. The second lesson is to note, when a problem is known and possible corrections have been identified, how few corrective options exist and how costly it is to implement them. Proposed fixes are, in the main, less than ideal; getting the plant back onstream is a driving force not to be overlooked. Combined, the lessons should be an impetus for developing complete specifications to begin with.

19.2 OBSTRUCTION OR LACK OF OBSTRUCTION TO FLOW

A representative fan performance curve was shown in Figure 1-1. Suppose system pressure loss increases beyond the value for which a fan was initially chosen. Operating against increased pressure, the fan will deliver less air and hence less cooling. The same is true of pumps. In subsequent sections of this chapter (Sections 19.3 to 19.14), always look for two conditions that are often the cause of underperformance. One is undersized ducts, pipes, or flow areas or other obstructions cause pumps and blowers to deliver less flow and hence lower performance. The second is *lack* of obstruction to flow. This condition can result from missing seal strips; missing welds in headers and pass partition plates; or loose tolerances, such as those between tubes and baffle holes or between shell and baffle diameters. These permit bypassing, reducing performance in a different way. Look for these conditions in the examples that follow and when faced with field problems.

In the discussion of racks in Part 3 of this book (Section 28.6), a serious problem occurs due to tight packing (Section 30.3), resulting in component overheating and failure. The cause is excessive obstruction to flow, much the same cause as cited in the examples that follow (see Figure 19-1, examples 1, 2, and 4).

19.3 UNDERSIZED SCREEN ROOM DUCTWORK

Most screen room electronic overheating problems (overheating in racks is a separate issue) occur because of reduced air flow to undersized (a) ducts or (b) screen rooms serving as ducts. Four designs are shown in Figures 19-2 to 19-5. Most overheating occurs in screen room designs similar to that shown in Figure 19-2. Its ECS is outside the room making duct lengths greater and the flow area to move air to under the floor is often minimal. Other designs shown are not as flow limiting. If flow were reversed (Figure 19-5) the area to the ECS would not be from one side but from two or more thus maximizing the approach area. Reducing flow to an ECS reduces its efficiency. It is usually beyond the control of the ECS manufacturer. Undersized screen room or duct systems, though unintentionally supplied, reduce flow to the ECS often causing its non-performance.

The "small ducting" conclusion is easily demonstrated by calculation. A unit similar to that shown in Figure 19-2 was to deliver 4000 cfm of air at a pressure drop of 0.5 in of water and was selected from a catalog (Table 19-1). The pressure drop available from the ECS is needed for moving air through the duct system. The blower must deliver another 0.5 in of water for flow across the filter and evaporator, or 1 in total. ECS manufacturers guarantee the output of their units. After installation the unit did not meet its guarantee, 10 tons. The question was, Why?

An ECS, installed in ducting to a screen room, was delivering much less cooling than its 10 ton design. The system is shown in Figure 19-2. It was

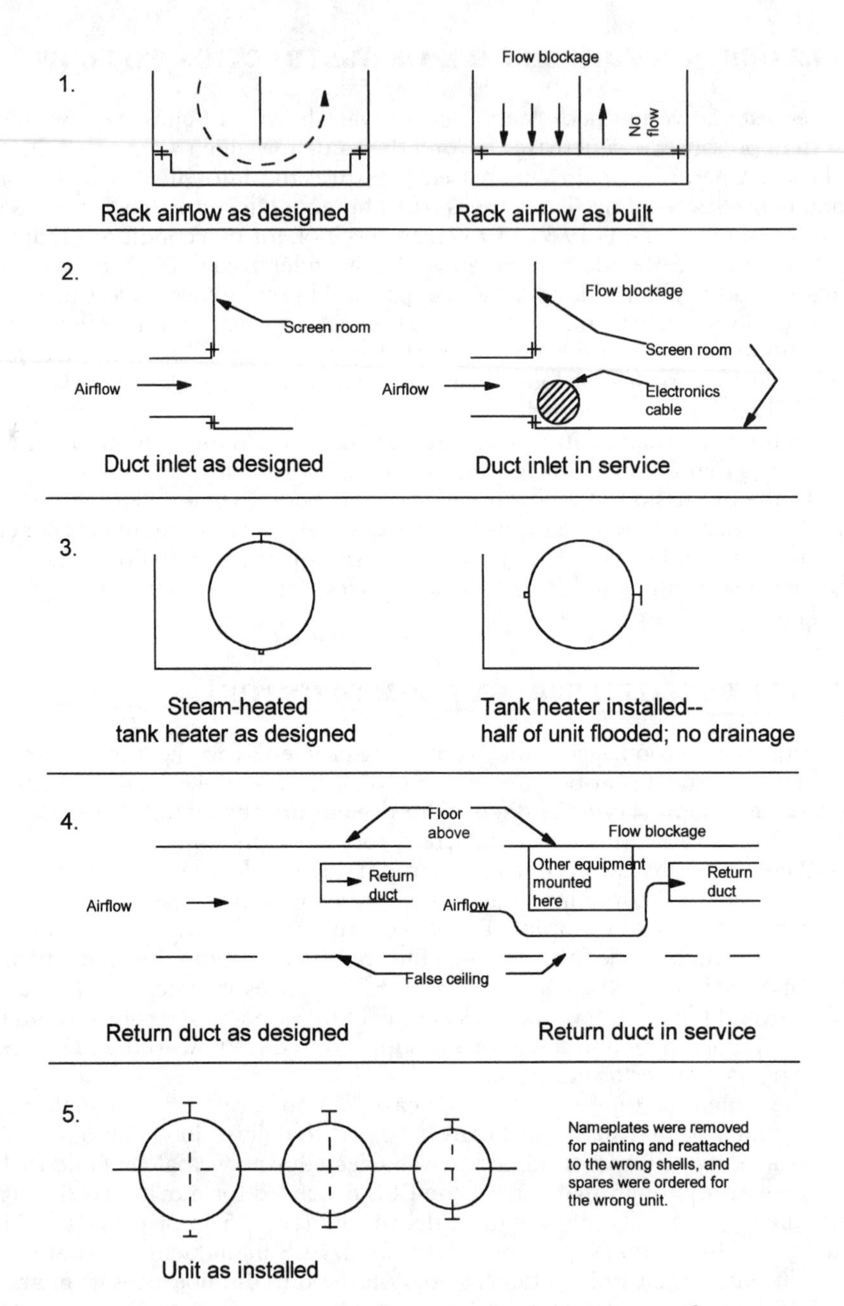

Figure 19-1. Various causes of failures, or why field trips are often necessary.

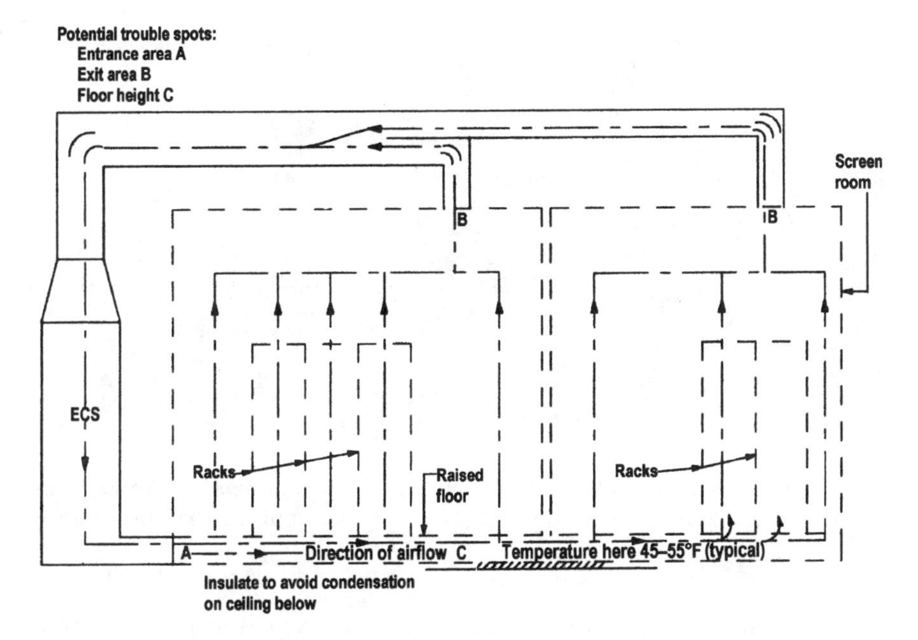

Figure 19-2. Diagram of basic ECS duct system for screen room. Cooling air enters from below a raised floor and returns through ceiling ducts to the ECS unit for recirculation.

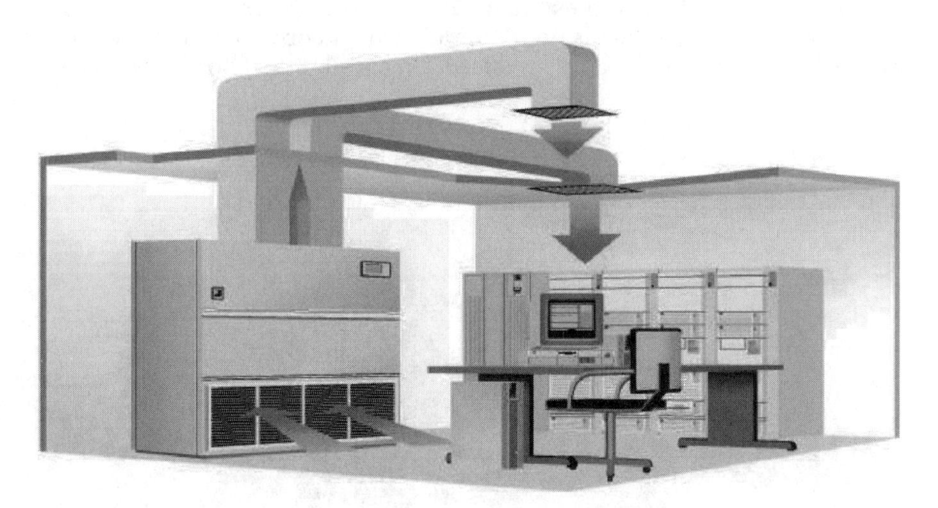

Figure 19-3. Screen room arrangement in which cooling air enters through ceiling ducts and returns to ECS unit at floor level. (Reprinted by permission of Liebert Corporation.)

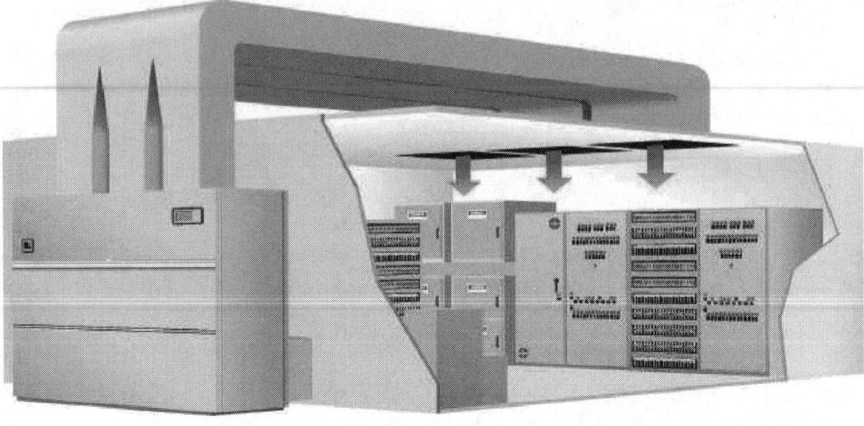

Figure 19-4. Screen room arrangement in which cooling air from ECS unit enters through ceiling ducts. (Reprinted by permission of Liebert Corporation.)

assembled in a building. Unfortunately, little space was available between the roof of the screen room and the building's ceiling, the ducting was lowered in elevation using short-radius elbows. This added to the system pressure loss.

The ducts carrying air leaving the screen room were larger in area than the elbow that supplied air to the area under the screen room floor. The vertical height of the floor opening is not 8-in; rather the floor was 8-in above grade. This effective opening height is reduced 2 inches by the floor thickness and

Figure 19-5. "Ductless" screen room arrangement. Cooling air enters room directly from ECS unit and returns beneath raised floor. (Reprinted by permission of Liebert Corporation.)

Table 19-1 Different Operating Conditions for an ECS Unit

Air-Cooled Data

Net capacity data Btu/h (kW), standard air volume and evaporator fan motor*

	DH/VH75A	DH/VH114A	DH/VH125A	DH/VHI99A	DH/VH245A	DH/VH290A	DH/VH380A
80°F DB, 67°F WB (26.7°C DB, 19.4°C WB) 50% RH							
Total	79,700 (23.4)	123,100 (36.0)	134,900 (39.5)	190,600 (55.9)	244,700 (71.7)	278,000 (81.5)	379,500 (111.1)
Sensible	64,600 (18.9)	93,000 (27.3)	115,800 (33.9)	169,200 (49.6)	208,200 (61.0)	237,400 (69.6)	312,600 (91.5)
75°F DB, 62.5°F WB (23.9°C DB, 16.9°C WB) 50% RH							
Total	74,400 (21.8)	114,200 (33.5)	125,500 (36.8)	178,000 (52.2)	227,400 (66.6)	259,300 (76.0)	353,400 (103.5)
Sensible	62,700 (18.4)	90,200 (26.4)	111,900 (32.8)	163,600 (47.9)	201,200 (59.0)	229,700 (67.3)	302,700 (88.6)
75°F DB, 61°F WB (23.9°C DB, 16.1°C WB) 45% RH							
Total	72,700 (21.3)	111,400 (32.6)	126,200 (37.0)	179,900 (52.7)	228,300 (66.9)	260,000 (76.2)	344,800 (100.0)
Sensible	67,000 (19.6)	96,000 (28.1)	126,200 (37.0)	179,900 (52.7)	228,300 (66.9)	260,000 (76.2)	323,600 (94.8)
72°F DB, 60°F WB (22.2°C DB, 15.5°C WB) 50% RH							
Total	71,400 (20.9)	109,100 (32.0)	120,100 (35.2)	170,400 (49.9)	217,600 (63.8)	248,400 (72.8)	338,800 (99.2)
Sensible	61,500 (18.0)	88,400 (25.9)	109,500 (32.1)	159,900 (46.9)	196,800 (57.7)	224,800 (65.9)	296,600 (86.9)
72°F DB, 58.6°F WB (22.2°C DB, 14.8°C WB) 45% RH							
Total	69,800 (20.5)	106,500 (31.2)	121,600 (35.6)	173,600 (50.9)	220,000 (64.5)	250,800 (73.5)	339,000 (99.3)
Sensible	65,500 (19.2)	93,800 (27.5)	121,600 (35.6)	173,600 (50.9)	220,000 (64.5)	250,800 (73.5)	339,000 (99.3)

*For optional fan motors deduct 2800 Btu/h per hp over standard motor.

Source: Reprinted by permission of Liebert Corporation.

another 2 inches by the lip at the screen room floor necessary to connect the elbow to the screen room. The elbow is limited in width by vertical columns that are part of the screen room structure. Hence, the elbow area is substantially undersized. (The simple solution, before assembly of the screen room, is to raise the floor several inches. After assembly, there is no easy solution). Another problem was that a large diameter cable was laid on the floor at this entrance connection (this error is diagrammed in Figure 19-1, example 2), further reducing the vertical opening. Calculations showed that the elbow width and floor height limitations added 0.25 in to the system pressure loss. Outlet elbows turns were more severe than normal so as to fit the available space between the building ceiling and screen room roof. Sharp turns added to duct pressure loss and were further tightened by expansion at the transition piece, at the floor elbow, and beneath the floor. The as-built losses were confirmed by calculation. A further change was that 30% efficient filters were replaced with HEPA filters. These have pressure losses of up to an inch, not the 0.2 in used in the calculations. The calculated as-built system pressure loss was 2.35 in, not 1 in.

At this condition the blower delivers 2600 cfm of air, not 4000. This reduces the system capacity from 10 tons at 4000 cfm to 6.5 tons at 2600 cfm. In other words the reduced cooling was caused by an undersized duct system that could not deliver 4000 cfm because of high duct pressure loss.

The foregoing example shows how lack of attention to detail can change a system's capacity. Field trips may be necessary to identify the problem because discrepancies or changes are frequently not documented.

The author was fortunate to see a nearly identical screen room and ECS installation at another location a few months later. In it the raised floor was 18 in high rather than 8 in, and the elbow from the ECS to the screen room was properly sized, as was the overhead duct. Short-radius elbows were not used, and 30% efficient filters rather than HEPAs were in place. Calculations for the earlier screen room applied and the installation was performing as guaranteed.

Conclusion: Confirm that units are constructed and installed as assumed in the calculations. The ECS was correct but an undersized duct system caused the drop in performance. In designs where the ECS unit is placed inside a room, as in Figure 19-5, the same problem could apply, that is, subminimum floor height, not enough clearance between the screen room and ceiling, and higher-grade filters but this rarely occurs in these designs. Placing the ECS in the room, however, has the disadvantage that it occupies space that could be used for racks or other purposes.

19.4 AIRFLOWS IN A/C AND ECS SYSTEMS

A rule of thumb in air conditioning is that 400 cfm of air equals one ton (12,000 Btu/h) of cooling. The ratio of the sensible heat to the total heat is called the *sensible heat ratio* (SHR). Its value is commonly given as 0.72.

An environmental control system (ECS) has an SHR (sensible heat ratio) of 0.86 to 0.95, not 0.72 because its makeup air is taken from the building, where moisture has been removed in the building A/C system. As a result ECS units are usually designed using higher airflows than A/C units, in the range of 550 to 650 cfm per ton of cooling. Some screen and clean room units have not performed because ducts were sized using the using the 400 cfm rule of thumb, not the higher airflows typical of ECS units. Others did not perform because makeup air was taken from outside the building rather than from inside. This increases the latent load on the evaporator, effectively making it undersized. When an ECS unit is not performing, one of these conditions may be the cause of the problem.

19.5 HEADER SIZE OF LUBE OIL AIR COOLER

A lube oil cooler that did not meet performance was described in Section 11.4. The cause of failure was similar to that of the undersized duct system described in Section 19.3 except that an undersized header and poor distribution were the cause of the problem. Undersized ducts, pipes, headers, or nozzles can reduce capacity. This applies to liquid and gas flows. Underperformance occurs most often in applications containing low-pressure liquid-gas mixtures or low-pressure gases. It seems the simplest solution is to increase the size of nozzles and headers. Header sections, where tube-to-tubesheet connections are made, are flat plates. Their thickness increases with increasing pressure. High-height headers cannot sustain the operating pressure of shorter-height headers, assuming the tubes of both are on the same pitch (refer to Table 4-2).

One way of limiting header size is to add more nozzles to headers. Nozzles' maximum size is limited by header width. An acceptable modification is to flatten a larger nozzle so that it is oval-shaped when welded to the header. If this is done, make sure that flow distribution is within acceptable limits. It happens this is the least costly way of proceeding. Economics governs. If the choice is thermally correct but not competitive, the selection is a failure. This applies to all coolers, not just lube oil units.

19.6 QUENCH OIL COOLER INLET DESIGN

Shell-and-tube quench oil coolers that did not meet performance were described in Section 15.2. The problem of performance was caused by the choice of tube layout, spacing, and nozzle design at the shell inlet. Flow was partially blocked by tubes positioned too close to the shell inlet nozzle as well as too close to one another. Both conditions prevented metal flakes from passing. These blocked the entrance area and the space between the tubes. The units would have met performance had the tubes been spaced further apart, on square pitch to simplify cleaning, and had the inlet nozzle had a dome. Cleaning lanes will make it necessary to furnish a larger shell diameter.

19.7 PLATE-AND-FRAME EXCHANGER

Water flowed in a plate-and-frame exchanger that was well designed. The water supply was turned off while repairs were made to an underground pipe. After the repairs were completed the water supply and exchanger were back onstream but no longer performed as designed. The exchanger was disassembled, and the cause of underperformance was immediately obvious. During pipe repairs large globs of clay entered the water supply and settled at the exchanger entrance. These blocked access to heat transfer surface. This example is much the same as the metal flaking problem described in Section 15.2. However, there were differences:

1. The clay blocked access to heat exchanger surface, effectively making it unavailable. This did not happen in the quench oil cooler example, because some flakes entered the bundle.
2. The plate-and-frame exchanger was easily disassembled and cleaned—an advantage of this construction. Disassembling shell-and-tube exchangers is not a realistic solution.
3. Once the plate-and-frame exchanger mud problem was corrected, the problem was over. The flaking condition in the quenching process shell-and-tube exchanger, however, is not realistically correctable without disassembling the unit to remove the metal flakes that settled in the interior of the tube bundle. Initially, metal filters could have been used to remove the flakes. However, air coolers with 1-in diameter tubes were ultimately chosen because their tube area is large enough to allow the metal flakes to pass, avoid the blockage problem, and they are easy to clean.

19.8 CUMULATIVE PRESSURE DROP

In trying to determine the cause of a problem, do not overlook the unintended consequences of an ordinary plant upgrade. It is very easy to use up or exceed the available pressure drop as more areas require service, or in situations where numerous units are used in series. The shortfalls could be called plant rather than exchanger problems, but the fact remains that they do occur, are exchanger-related, and result in reduced plant capacity. The deficiencies described in Sections 19.9 ("Chiller Plant") and 19.10 ("Refinery") are very similar in essence though not in their details. Furthermore, these problems are not that different from the lube oil cooler problem (Section 11.4) and the screen room ductwork problem (Section 19.3).

19.9 CHILLER PLANT

A chiller was designed to deliver cold water to 10 floors of a building. At start-up and when new, the plant performed as designed. A while later, facilities on the fourth and seventh floors were upgraded, requiring added cooling. Soon thereafter, similar situations were faced on the third and eighth floors, each requiring additional cooling. Thus, more water had to pass through the line between the basement and eighth floors, increasing the pressure drop in this line. The result was that less pressure was available at the 10th floor. The 10th floor air-handling unit was designed for a 10 psi pressure loss, but the line pressure at this position was now 5 psi. Under these conditions, flow to the exchanger was reduced and the air handler would not meet performance. A simple though costly solution would be to add a booster pump to increase the line pressure to the 10th floor. An argument could be made that this situation is another variation on the flow obstruction problem of Section 19.3.

19.10 REFINERY

In refineries, pressure-drop crises can affect both shell-and-tube and double-pipe exchangers. There are conditions where it is not possible to have pressure in the tubes without having it in the shells. The pressure to both streams is delivered by the same pump. The typical application has several exchangers in series, and typically the designer allocates 10 psi of pressure loss to each side of each exchanger. For four units in series the allowable pressure drop is then 40 psi through the tubes and 40 psi through the shells or 80 psi total. The exchanger manufacturer, however, is nearly always allowed to apportion this total drop in any way that results in the lowest-cost bank of exchangers that provides the desired heat transfer. As an example, an exchanger may have a pressure drop of 7 psi in the tubes and 8 psi in the shell. The unit following may have a pressure drop of 13 psi in the tubes and 12 psi in the shell. In this case, both streams have a pressure drop of 20 psi in each stream of two exchangers in series, meeting the allowable pressure loss. Similar reasoning applies to any other group of exchangers in series. The point is that when the option of allocating pressure drop exists, more pressure drop will be used than if the designer is limited to 10 psi per stream per exchanger. This reduces the pressure drop available for future modifications and can result in using more pressure drop than available. As an example, the dimensions of an 11 BWG average wall tube should be used for a 12 BWG minimum wall tube when the pressure drop calculation is made (Table 2-4) not a 1-in 12 BWG theoretical tube.

If a plant upgrade is made that results in exceeding the allowable pressure drop, the pump will not deliver the required flow and the performance of the exchangers will decrease. This can be a major problem because it can reduce a refinery's output by several percent. Adding a pump in parallel, though costly, may be the best remedy. Note the similarity of this problem and that of undersized ducting described in Section 19.3.

Warning to heat exchanger manufacturers: When allocating pressure drop over a series of exchangers, it is easy to use the pressure drop available. Check your numbers closely. If you guarantee a total pressure drop of 80 psi and the actual is nearer 100, the plant may underperform.

19.11 STARVING FANS

A common cause of thermally correct units not performing is preventing cool air from reaching their fan or blower. In nearly every case the units delivered by the manufacturer will deliver the cooling air required. The problem occurs due to site details that prevent sufficient air from reaching the fan or blower.

In Section 19.3 a blower did not deliver the quantity of air for which it was designed because its supply and return ducts were undersized. When propeller fans are used they are not normally supplied with ducts but they may face a similar problem, that is, insufficient air to the fan. To prevent this from happening it is recommended that the air supply area to the fan be 125% of the fan area.

Most fans furnished with air coolers provide sufficient air flow for units to perform. Problems surface based on specific conditions at the installation space provided. Air coolers installed near buildings often have their air supply blocked by these buildings preventing air from reaching the fan(s). This effectively blocks flow to one or more sides of a cooler. In some installations fan air may be available from one direction only. When the installed air supply flow area to the fan(s) is less than the fan(s) area (not the supply flow area provided with the unit), fans are said to be starved.

The recommended 25% of fan area over that of the actual fan area is not realistically a factor of safety but is one of convenience. The gross area to the fan is easily measured but is partially blocked by columns, portions of fan rings, or by structural members. Rather than spending a great deal of time calculating these obstructions, the 25% rule of thumb usually suffices. The author cannot recall a process or A/C unit propeller fan that was limited by a minimum outlet air flow area (not true of duct blowers or electronic cooling fans). This does not mean that fans were never undersized which is not the same as starving a fan.

A word of caution about two types of installations. In process coolers that are provided with recirculation chambers, the added structural members or needs (louvers, sliding doors, access doors, removable panels, and cold weather sealing devices) all occupy space that is to some degree in the air flow path. Make certain that these, as a group, do not create a "starving the fan" situation.

The second application applies to fans that move the air in electronic boxes. It applies whether fans are installed in forced or induced draft. The blowers are small in size, usually 2-, 3-, 4-, or 6-in diameter. When mounted in an electronic box and in a rack, the fans should have space to allow the inlet air to enter the box unimpeded; outlet air should also have an unobstructed exit path. Unfortunately, electronic boxes are often mounted too close to rack structural members or other electronic boxes preventing them from receiving the cooling air they need or preventing warmed air from exiting the box (Figure 28-2). Both are "starving the fan" situations. This is one problem area of tight packaging of electronic equipment in racks; the other is providing too little flow area for air to pass. (See also Sections 29.9 and 30.3.)

19.12 FLOW RESTRICTIONS

Consider all the previous topics and examples in this chapter. In the screen room application, duct restrictions and filter resistance reduced flow, causing underperformance. The ECS ducting was undersized because the 400 cfm rule-of-thumb flow was used rather than the actual 550 cfm. In the lube oil cooler, the obstruction was an undersized header, while the quench oil cooler had a partially blocked nozzle, an undersized flow area into the bundle, and tightly spaced tubes and lacked a dome area. The problem in the plate-and-frame unit was a one-time blockage that could not have been anticipated. The chiller system (not the chiller) did not perform because water flow to the lower floors was increased, reducing the pressure available at the upper floors. In the refinery, a bank of exchangers was chosen that exceeded the available pressure drop. In all cases, including starving fans, flow restrictions were the cause of reduced performance. Look for conditions like these when trying to determine why units do not perform. The author's experience is that the most frequent cause of underperformance is restricted flow.

Now refer to Part 3 of this book, on electronics cooling. The examples given illustrate that flow restrictions also cause underperformance in this industry. In the previous examples obstructions to flow were the cause of reduced capacity. Most were caused by undersized flow areas in ducts, piping, manifolds, or headers resulting in high local pressure losses. Relocating the lube oil cooler inlet nozzle was not approved by engineering and created a similar problem. A frequent cause of underperformance is approval of changes by unqualified personnel for expediency. No equipment inspector is likely to be aware of blockage problems while witnessing equipment being built, but engineers of the buyer's or seller's organization should be ready to question conditions such as the lack of a dome area or to look for high cumulative pressure losses that exceed the allowable.

19.13 AVOIDING BYPASSING: A CHECKLIST FOR INSPECTORS

In the following examples underperformance is the result of conditions that are the responsibility of heat exchanger manufacturers. They include details that buyer's inspectors should look for during shop visits and inspections.

19.13.1 Pass Partition Plate Welds

Figure 19-6 shows the cross section of an air cooler header. At the header end plates, a continuous weld should be provided to join end and pass partition plates inside the exchanger to prevent flow from short-circuiting. The potential for bypass exists at the end plates and at the pass partition plate in the inlet header of a two-pass unit. On other projects, similar conditions may exist for a four-pass header, except that there will be two pass partition

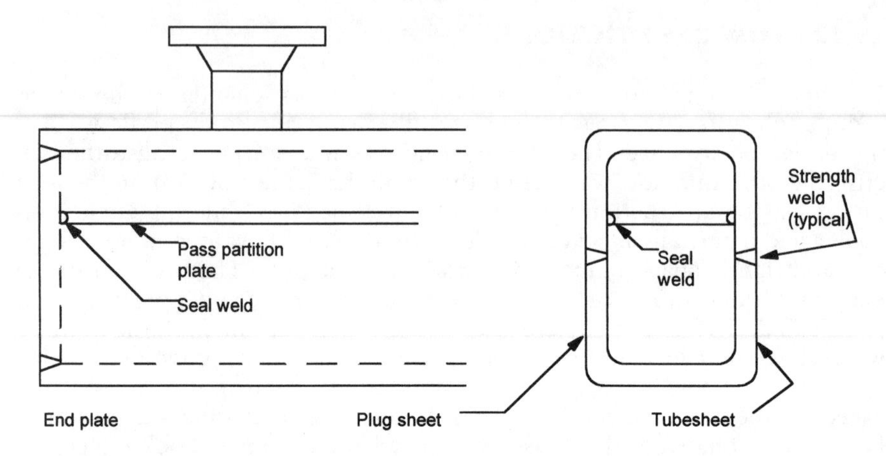

Figure 19-6. Header welds to prevent bypassing.

plates in the inlet header and one in the floating header. Similar reasoning applies to other pass arrangements. Depending on pressure, these may be seal or strength welds. In any event, their omission will cause fluids to short-circuit, bypassing surface; and will reduce flow through the exchanger and cause a drop in capacity. (Keep in mind that omitting welds also causes bypassing in other kinds of exchangers.)

Weld omissions should be spotted by buyer's and seller's inspectors. Once corrective seals have been made, the exchangers should meet performance. While reduced flow caused by bypassing results in a drop in capacity, the cause is difficult to locate because, once an exchanger is assembled, welds are not normally visible from the outside. To confirm that a weld is missing may require that an exchanger or plant be shut down. If evidence of a weld can be seen from outside an exchanger, however, it is not likely that this will be the cause of reduced capacity.

19.13.2 Seal Strips

Air Coolers

Figure 19-7 shows a cross section of some rows of spiral finned tubes near an air cooler side channel. The rows are staggered. There are larger flow areas in some rows than others. For this reason seal strips are used that force air across the tubes to prevent bypassing.

Consider the case of not installing seal strips. At the end of a row (the last two or three tubes) air passes over half as many tubes as at other comparable clusters along the row. Therefore, in a four-layer section, the pressure drop at the end of the row will be half that at other locations along the row, because only two tube rows out of the four are crossed. Air takes the path of

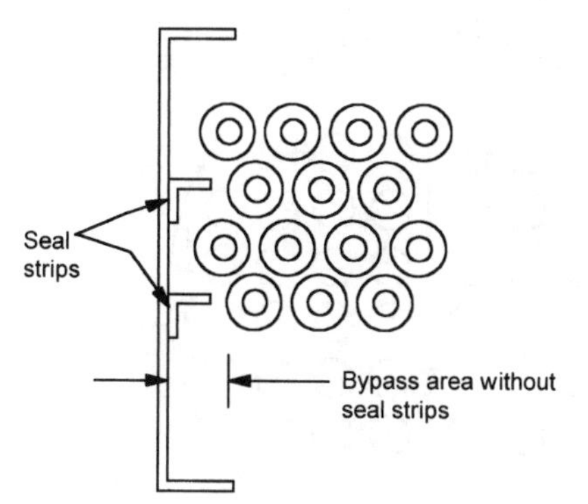

Figure 19-7. Cross section of rows of tubes near air cooler side channel, with seal strips. Without seal strips, bypassing would occur through area indicated.

least resistance (greatest open area), and without seal strips, ends of rows are where the flow area is greatest. It is not unusual to lose 30% of an air cooler's capacity to short-circuiting at these locations. Therefore, seal strips are a must.

Note to inspectors: Confirm that seal strips have been installed.

Shell-and-Tube Exchangers

Seal strips are needed in most shell-and-tube units. A cross section, with and without bypass-preventing devices, is shown in Figure 19-8. The same rules apply as for air coolers. Seal strips direct flow through the bundle. Fluids take the path of least resistance (around the bundle) unless some kind of blockage to flow is installed. If fluids take the bypass route, they avoid heating or cooling surface. This causes underperformance. It should be noted that some isothermal condenser applications do not require seal strips because the condensing driving force is large; see Section 18.2. On the other hand, seal strips are a must in mixed-flow condensers.

Figure 19.8(b) shows several designs that serve the same purpose as seal strips. Their advantage is often ease of installation particularly when they are located in the internal area of a bundle. The flat seal strip is an anti-bypass device almost always positioned at the periphery of a bundle. Short-cut passes through bundles, created by a need for pass partition plates, must also have some form of seal strip to direct flow into the bundle. Consider tie rods which are needed to hold baffles in position. Why not have them serve a dual purpose? If located near the outer edge of a bundle, they serve almost like another tube to direct flow into the bundle. A modification of this is the winged spacer. Yet another sealing device is large diameter tubes positioned opposite a pass partition plate in most cases. They serve the purpose of seal strips and are much easier to install and hold in position than flat strips

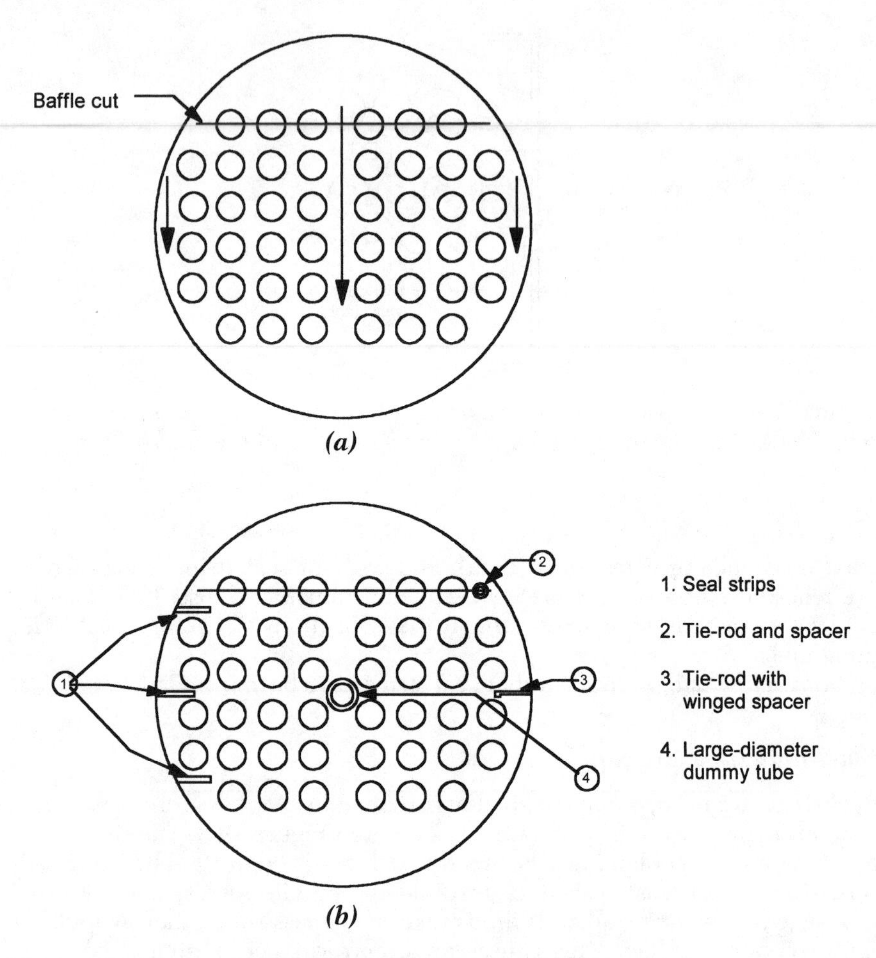

Figure 19-8. Cross section of tube layout in shell-and-tube exchanger (*a*) showing bypass lanes and (*b*) fitted with various devices for preventing bypass.

placed at the same locations. If the bypass path is narrow, tie rods positioned along it serve the same purpose.

Note to inspectors: Confirm that seal strips have been installed where needed.

19.14 BAFFLES

Baffles are used in shell-and-tube exchangers to support tubes and minimize vibration, but their main purpose is to direct flow across and through a bundle.

19.14.1 Baffle Outside Diameter

The baffle OD is ideally the shell ID. However, space must be provided to remove and install bundles without interference. Also, a minimum out-of-roundness exists because shells and baffle segments are not perfectly round. When the shell ID and baffle OD are within tolerance (standards exist), exchangers should perform as guaranteed. When the looseness of fit exceeds these tolerances, a horizontally installed bundle (with its baffles) will rest on the shell ID, creating an open space at the top of the shell, shaped much like a one-eighth moon. This is a bypass route and is to be avoided. On the other hand, too large a baffle OD makes it difficult to remove or install a bundle.

Note to inspectors: Confirm that clearances between the shell ID and baffle OD are within tolerance.

19.14.2 Baffle Hole Tolerance

The other fit that is required of baffles in shell-and-tube exchangers is between the ID of the baffle hole and the OD of the tube. Some clearance has to exist in order for tubes to pass during assembly. Clearances have been established based on experience.

Heat transfer in crossflow is more efficient than in long flow. If the baffle holes are significantly larger than the tube OD, more fluid takes the longitudinal flow path. When this happens, the shell-side heat transfer rate and the overall U are reduced, sometimes causing underperformance. No fix exists for this condition other than adding surface in another exchanger or replacing the bundle. Hopefully, when needed, space will be available for the former to be accomplished. If baffle holes are oversized, vibration often becomes a problem and is costly to correct.

Inspectors: Confirm that baffle hole tolerances are met.

19.14.3 Longitudinal Baffle

A low-cost alternate for two shell-and-tube units in series is one shell with a longitudinal baffle. Typical TEMA nomenclatures are AFL, BFN, or BFS (see Figure 3-1). The advantage of this construction can be an improved LMTD along with eliminating one shell; one shell with a longitudinal baffle usually costs less than two smaller shells. Should bypass occur, it will likely be because the integrity of the spring seal between the longitudinal baffle and shell is no longer functioning as designed, with results similar to what happens when a weld is omitted in a pass partition plate, as described in Section 19.13. The difference is that bypass takes place on the shell side. In this construction, shell inlet and outlet nozzles are opposite each other. The pressure differential across the longitudinal baffle at the nozzles is almost equal to the pressure drop through the shell. On a 2-ft by 2-ft section of baffle, assuming a 10 psi pressure drop through the exchanger, the load on the baffle will be:

$$(2)(2)(144)(10) = 5760 \text{ lb}$$

A load of this magnitude tends to warp longitudinal baffles. It is difficult to maintain the seal between baffle and shell following disassembly and reassembly of the unit. Bypassing is apt to occur where the differential pressure is greatest, near the tubesheet. A leak will cause some fluids to bypass surface, resulting in a loss of capacity. A good rule is to minimize pressure drop in a two-pass shell to reduce the load on the longitudinal baffle.

19.15 OTHER CAUSES OF UNDERPERFORMANCE

The skills needed to size an exchanger are not the same skills as those needed to determine why an exchanger is not performing and then take corrective action in the field. Maintaining heat exchanger performance requires not only troubleshooting skills, but also awareness of when and how to make adjustments. For example, process and power plant systems have to be taken off-line when changes are made, but A/C system changes can be made without shutting down. If insufficient air is being delivered to a room, closing a damper can isolate the affected area, allowing duct and outlet changes to be made while the system as a whole remains on-line; after completion the damper is opened, allowing air to flow. (An experienced engineer or troubleshooter would know that replacing an air filter with a more efficient one can result in a large reduction in the A/C unit's performance.)

Here are additional examples that have caused systems to underperform. In the usual case problems like these are not recognized by the engineer until trouble surfaces. The examples should give engineers an idea of problems to expect.

19.15.1 River Water Temperature

A bank of generator coolers was placed in service and performed for 30 years. One summer their output dropped. When the installation was originally designed the cooling source was river water at 67°F. Over the years the river's temperature rose more than 13°F because of thermal discharges from new industrial plants built further upstream. The warmed river changed the temperature at the exchanger's inlet.

Little can be done to lower the river water temperature in the short term. Hence, surface must be added, or the unit will continue to underperform. Typically, air or hydrogen coolers occupy volumes in four quadrants of a generator. Seldom does the surface needed occupy the available volume. The preferred solution is to furnish bundles that fit the available space (more fins per inch and more tubes usually suffice). The plant can remain onstream while new bundles are built. Once the bundles have been delivered on-site, a generator can be turned off during periods of low power demand (such as weekends) to replace bundle(s) with minimum downtime.

19.15.2 Channeling

When two or more tubes or streams are in parallel in the transition or laminar region and are being cooled, one stream will cool slightly more rapidly than the other. Eventually it will become colder and more viscous, reaching a point that fluid in these tubes will no longer flow. When this happens, flow in the other parallel stream increases. This is called channeling. Because of increased flow in some tubes and because part of the cooling surface is no longer functional, the portion of the exchanger where the fluid continues to flow will not provide the cooling desired. The possibility of channeling should be addressed at the time the initial selection is made.

There are three commonly used methods of reducing the chances of this problem occurring. If you know the likelihood of channelling in your situation is minimal, select a unit for parallel flow operation. The selection will be larger in size than if a counterflow exchanger were chosen. Note in Figure 2.1 that the benefit of parallel flow is that it limits the amount of cooling that can be attained in an exchanger which is why it can be used to advantage in this case.

A second method is to cool the product keeping the flow in a single stream using large-diameter pipes or tubes in series. They are installed near the base of a tank that is filled with water and kept warm enough to prevent the no-flow condition from occurring.

The third method is to use heated air or water to avoid the overcooling that causes channelling. It differs from the second method in that many streams in parallel can be used.

19.15.3 Rack Drawers

Racks are designed for mounting electronic equipment. Electronic components are packaged in modules for ease of handling and to save space. Rack panels act as the sides of a duct and contain the air that is to keep the equipment cool. They make up part of the room air-conditioning system, but changes can be made to them without turning the A/C system off. When a packaged unit is no longer needed, it is removed, leaving an empty space in the rack. A potential (emphasis on *potential*) error is to install a drawer in this space for storing equipment or manuals. A drawer or a flat plate can block airflow to nearby equipment, causing it to overheat. This condition does not apply to all drawers, but it does occur in about 1 of 5 cases. One suggestion is to have the drawer perforated, allowing air to pass. Even then, the contents of the drawer can act as a flat plate: manuals, for instance, lying on perforated plates will still block flow. In roughly 80% of cases an effective modification is to design the drawer so that it is not longer than three-fourths the depth of the rack. The open space beyond the drawer will usually allow sufficient air to pass.

Many drawers are installed without considering A/C requirements. While their use does not interfere with the operation of the main A/C system, blocking air to electronic devices sometimes causes them to fail. The usual maximum temperature of electronic equipment is 104°F (40°C). As a comparison, process exchangers may transfer millions of Btu per hour, while rack cooling loads may be as low as 50 to 300 Btu/h. Although the cooling loads are small, the heat is enough to cause equipment to fail.

19.15.4 Rack Plate or Flow Channel

Figure 19-1, example 1, shows a rack and panel designed for reversing the direction of airflow. Without thinking of A/C needs and for simplicity, unauthorized personnel substituted a flat plate at this juncture. Unfortunately this does not allow air to reverse direction. The flat plate eliminates needed cooling, thus causing equipment to fail. It is typical of unauthorized changes that cause failures. A good idea is to provide markings on panels such as this one indicating that they are not to be removed.

19.15.5 Tank Heater

A tank heater that did not perform is shown in Figure 19-1, example 3. Steam for heating the product should have entered at the top of the channel, flowed through tubes, and condensed. Water was to drain through a nozzle at the bottom of the channel. The unit was improperly installed with the centerline of the nozzles located in the horizontal plane. This arrangement flooded half the unit, leaving only half the exchanger for heating. Improper installation reduced the unit's performance. To correct the error, assuming no shutoff valve exists, requires draining the tank, deciding where the product will be stored in the interim, deciding if the viscous liquid can be drained in a low-temperature environment (usually it can't), and then rotating the exchanger.

Some errors are difficult to correct. The point is that when submitting drawings, include all details, as few can be overlooked. An error as simple as this is a good example.

19.15.6 Sizing Errors

Thanks largely to computers, few exchangers today are sized incorrectly. This does not mean none. Sizing-related underperformance has resulted from various kinds of oversights:

1. Fluid properties were published in error and later corrected but the older data were still used to size equipment.
2. The flow changed from laminar to turbulent in the exchanger but the effect on heat transfer rate was not considered in sizing equipment.
3. Films form when hydrocarbons are condensed but not when steam is condensed. Which condition applies to the exchanger being sized?
4. The condensing load is overlooked when air is cooled.

5. Properties change when two or more fluids are mixed. Examples are adding turpentine to paint or steam-heating vegetable oils by direct contact.
6. In A/C systems, the return of oil by gravity to the compressor must not be overlooked. The unit supplied must be capable of freely draining, otherwise, failures will result.

19.16 SUMMARY: CAUSES AND COST OF FAILURES

Several thoughts emerge from a review of the many examples presented in this chapter. A recurring theme is that the causes of underperformance are often invisible from outside an exchanger without disassembling the unit. Consider the omission of a pass partition seal weld as shown in Figure 19-7 or of side seal strips like those shown in Figure 19-8. A similarly hidden condition is too large a clearance between tubes and baffle holes in a shell-and-tube exchanger. With any of these conditions present, underperformance will not become known until units are operating.

An interesting situation was the cumulative addition of pressure drop described in Section 19.10. In this example, the exchanger may not necessarily be underperforming, but the plant will be.

Aside from the causes of underperformance described in this chapter, there are other reasons why thermally correct exchangers can give dissatisfaction:

1. Units fail rapidly if the materials selected are not compatible with the process.
2. Servicing may be unnecessarily inconvenient. Many a maintenance crew has used expletives to deride engineers for complicated gasketing.
3. Because a problem was not thought through, units have been delivered that required mechanical cleaning but were incapable of it.
4. The possibility of recirculation (discussed in Chapter 20) was not considered when a unit was sized.

The cost of correcting an underperforming exchanger takes different forms. If corrections have to be made in a refinery, the fix usually requires that the plant be shut down, a costly step, as several days production will be lost. Shutting down a plant is not an easy decision. Even where a major shutdown is not a necessity, a great deal of planning is required to make the needed corrections while the plant is onstream or at least to keep downtime to a minimum. Compare this with how simple the "fix" would be if it were made when the unit was being manufactured.

Because of consequential costs associated with other equipment, some of the problems noted in this chapter were not corrected. The cost of making them exceeded the cost of the exchanger. In many cases there are no reasonable fixes without excessive cost in time and money. Further, the rating engineer is not normally aware of these conditions when selecting equipment. It is reasonable to conclude that these problems could have been avoided had the personnel listed in Section 1.2 done their job more thoroughly.

Sometimes the rating engineer has taken into consideration all requirements given in the specifications, has made a good selection, and has seen to it that the unit was built as designed, but is later advised that it is not performing as expected. Under these circumstances, Table 1-9 ("Some Causes of Process Units' Not Meeting Requirements") in Chapter 1 can be helpful. This table is a comprehensive list of possible causes of underperformance.

Chapter 20
Recirculation

Air recirculation and starving fans are addressed together since their correction nearly always requires a structural change. Once built, structures can be difficult and costly to change.

20.1 AIR COOLERS

Recirculation is not unusual when air or water is the coolant. The goal is to control it. Consider a forced-draft air cooler. Ambient air is heated as it passes through the coil. When leaving, the air at the cooler periphery can be drawn into the suction path to again be part of the entering cooling stream. This small stream of air will, after mixing, raise the entering ambient temperature to the coil, reduce the LMTD (log mean temperature difference), and require that a larger exchanger be provided. Similar conditions occur when water is the coolant.

Designs exist that minimize and effectively eliminate the problem of recirculation when air is the coolant. A good example is to use the induced-draft cooler design. In this the fan is positioned to direct exiting air upward at high velocity in a direction away from the entering inlet air. This reduces the chance of pulling a large quantity of air into the inlet stream. In the forced-draft design, recirculation can be minimized by adding a wind fence, should one be necessary. The fence surrounds a bay and contains the air so that it can exit only from an opening 4 to 5 ft above the coil. Distance then separates the streams, minimizing the chance that recirculation will occur (see Figure 6-1).

Consider once more the bank of exchangers shown in Figure 6-1 with the prevailing wind direction along the bay widths (forced or induced). Warm air leaving bays 1 and 2 will be blown toward bays 3 and 4. This warm air will have a component that, at times, will recirculate through bays 3 and 4. At other times, when the wind blows in the opposite direction, bays 1 and 2 will experience the same condition. In this situation, it is good practice to raise the design ambient temperature 1 or 2°F to account for recirculation.

20.2 STARVING FANS

The subject of starving fans is best illustrated by example. Suppose the need is for a process cooler 8 ft wide with 30-ft tubes. Two fans in one bay will deliver the desired cooling air. The following geometry then exists:

- The coil face area is 8 ft by 30 ft, or 240 ft^2. Its plenum height is 5 feet.
- The fan area is 50.3 ft^2 per fan (for 8-ft-diameter fans), or 100.5 ft^2 for two.
- The fan coverage is 100.5/240, or 41.9%. Good design practice is that this proportion should be a minimum of 40%. (Note that 7-ft-diameter fans would not meet this requirement. Also, if the fans are direct-connected with a minimum-height plenum, a higher percentage of coverage is required. See Section 6.1.)

Now consider the unit standing alone with its fan deck mounted 4 ft above grade. The vertical area available for air to reach either fan is:

$$(4 \times 15) + (4 \times 8) + (4 \times 15) = 152 \text{ ft}^2$$

The 8-ft-diameter fan area is 50.3 ft^2. Thus a flow area of 3 times the area of the fan is provided to feed air to the fan (152/50.3). The same unit when positioned near a building cannot realistically pull ambient air along the sides adjacent to the building. The area for flow may be from one side only and would be 60 ft^2 (4 × 15). (For simplicity, the deduction for area blocked by columns was not included in this calculation, but in actual design it should be.) Hence, the area for airflow relative to the fan area is (60/50.3) or 1.19. Ideally, this factor should be 1.25 minimum. Lengthening the structure columns by 1 ft or more will solve the problem.

It is relatively common in refinery and natural gas services to locate the fan deck approximately 6 ft above grade to service drives. Here is a reason. Suppose a forced-draft cooler is ground mounted and located in a bank of exchangers but not in an outside bay. (See bays 2 and 3 of Figure 6-1.) The design is such that air can enter at one end only, and the flow area to the fan is 8 ft by 6 ft, or 48 ft^2. The fan area is 50.3 ft^2. With these dimensions, the supply area is smaller than the recommended 1.25 times the area of the fan. Built this way the construction is said to starve the fan. In this ground mounted case the fan deck and coil should be elevated from 6 to 8 ft above grade to avoid this problem. For piperack mounted units, the area beneath the coils will be available to supply air to the fans. Starving a fan almost never occurs in rack mounted units. Before finalizing the placement of air coolers in a refinery or chemical plant, determine if there are other coolers either at the same elevation or above or below the units being considered. Hot air from these units has the same effect as recirculated air, the only difference being that the unwanted heat is from a different source. Review and make the needed corrections.

20.3 BUILDING-AIR RECIRCULATION

Recirculation of air to and from a building is a problem analogous to recirculation to and from an air cooler. Recirculation originates outside the building and is influenced by inlet and outlet duct locations. Therefore, preventing the problem is largely a matter of good placement of air inlets and outlets to ensure a supply of fresh air. Recirculated air is not fresh air. It should be avoided like other undesirable contaminants; in particular, it results in an increased cooling load on the evaporator.

Locate the spent air exhaust as far as practical from the air inlet so the effect of one on the other is minimal. For aesthetic purposes, avoid locating inlets or outlets on the front of a building. Do not locate an air inlet near:

- A paint spray exhaust
- A loading dock, from which excessive vehicle exhaust may enter the building
- An air outlet from a nearby plant
- Any source of odors

If some recirculation cannot be avoided, increase the quantity of makeup air to account for the fresh air displaced by recirculated air.

20.4 DUCT DESIGN

Recirculation entails mixing of hot and cold air. Heat transfer equations can be used to calculate the mix temperature where these streams join. Be aware that a calculated mix temperature above 32°F is not a guarantee that freezing will not occur. When streams meet, complete mixing does not occur immediately. It takes time and eddies (turning losses, particularly) to complete the mixing process. Mixing in ducts is usually complete after the combined stream experiences one turning loss plus a travel distance of seven equivalent duct diameters. Said another way, if outside air is a few degrees Fahrenheit below freezing and these restrictions are not met, a high percentage of air crossing the coil will be warm enough due to mixing to avoid freezing but a remnant of cold air flowing along one side of the duct will be enough to cause localized freezing and coil failure, usually requiring coil replacement.

20.5 WATER RECIRCULATION

Air recirculation is more common than water recirculation. If plant water comes from a river, the inlet is located on the upstream side of the river and the discharge on the downstream. River flow prevents recirculation from occurring. Inlets and outlets are spaced as far apart as practical.

An exchanger's cooling capacity can be reduced by factors that occur away from the facility. A typical case is a new industrial facility built further up the river whose output raises the river water temperature. See the discussion of river water temperature in Section 19.15. Plants are frequently exposed to a thermal discharge from neighboring facilities.

Assume recirculation results in an exchanger that is too small to perform the desired cooling. The question is, What should be done? In general, this problem will not be known until a plant's output is less than design conditions. The engineer is called and asked what can be done to fix the problem. Under these conditions, what is needed is not the best but the quickest solution that will get the plant onstream at full capacity with minimum disruption. Here are some suggestions.

1. Replace the tube bundle with one that uses 3/4- rather than 1-in-OD tubes. This will increase the surface in the shell by between 15 and 35% depending on the shell diameter. If this is done it is likely the tube side pass arrangement will need to be modified. This work should be scheduled for completion during the time the plant is down for bundle replacement.
2. Replace the tube bundle with one that uses lowfin tubes. (These tubes are usually not as readily available as bare tubes.) The tube-side flow area will be reduced possibly resulting in a need for a tube-side channel modification, see point 1 above.
3. Add one or more units in parallel.
4. When more cooling is needed for motors or generators cooled by air or hydrogen, consider replacing the bundles with units having more fins per inch and/or more tubes. Units that have been in service a long time often have fewer fins per inch.
5. Another option for air- or hydrogen-cooled units is to add a one-layer section on top of an existing section. Check that the existing fan will deliver the required air or air/hydrogen mixture.
6. Replace a tube having longitudinal steel fins with one having aluminum or copper fins.

Nearly all the suggestions for getting a plant back on-line are directed toward placing more surface in an existing volume, because this minimizes plant changes. Replacing a bundle only in a shell-and-tube unit requires additional work. Although tube nozzles must be disconnected and reconnected, shell nozzles will not be disturbed. No change to the exchanger's mounting is needed. The needed work will be mostly on the new bundle, especially if a new tube diameter or the same tube diameter with lowfins is chosen. A new bundle drawing must be made, new tube sheets and baffles must be drilled, and the channel pass arrangement may have to be modified. When the bundle is completed, installation can begin. Under urgent conditions bundles can usually be built in two weeks. Complete units take longer. This case demonstrates the kind of reasoning needed to get a plant back online quickly. It does not emphasize exchanger sizing as much as working with what exists and adapting to current needs.

Plants built on lakes do not have the advantage of discharging water downstream. Recirculation problems may be introduced because winds or currents can direct discharged water back to the plant's inlet. To counteract this, inlets and outlets are spaced a good distance apart. Naturally, the size of the lake has a large effect on the potential for recirculation. In any event, this is a factor to be considered in establishing plant water design temperature.

20.6 CONTROLLED RECIRCULATION

In air coolers, cold air can freeze water in tubes. A similar problem occurs in other applications when a product is cooled below its pour point thus plugging tubes and stopping flow (refer to the discussion of channeling in Section 19.15). Both conditions should be avoided. One way is to limit temperature drop so that a product does not experience the cold temperatures that cause problems. Another is to build a recirculation or freeze protection system on the air cooler. The heat source is the warm air off the coil. The idea is to recirculate some of this warm air with incoming cold air. Mixing begins when warm and cold streams meet. Warmed ambient air prevents the tube temperature from falling to levels that cause freezing or pour point problems. For two-fan air coolers with a recirculating air system, both fans should run to provide good mixing. Fans, at times, can be run at half speed when this feature is provided. Positive mixing is essential to prevent freezing. In this system, the product temperature is controlled by the fan blade pitch. Louver positions are controlled by the air mix temperature at the inlet to the coil. This design has had good success in operation.

A word of caution should be added. The system must be checked to confirm that fans are not starved during any season or operating scenario. It may require doors that are opened in summer or another means of ensuring sufficient air to each fan.

20.7 RECIRCULATION HINTS

Controlled air cooler recirculation systems have been designed, built, and operate well under various scenarios if the operation is continuous. Problems surface because of startup or shutdown conditions. Nearly all recirculation problems are operating problems. If a facility is shut down for any reason , or if it anticipated that at some future time it may be forced to shut down, or at startup, the immediate conditions that may have to be faced are freezing and solidification. Some proposed measures to correct these problems are given in Chapter 12.

Chapter 21
Shipping and Handling

At first glance the subjects of shipping and handling do not appear to be part of the selection process. For smaller units they are often not a factor. In larger units a closer look at the details shows their significance, which is best illustrated by examples.

21.1 SHOCK LOAD

Consider a large shell-and-tube unit shipped by rail and anchored to a railroad car bed with its tubes parallel to the tracks. There is a troublesome shock load condition that can damage the exchanger that occurs when railroad cars are assembled to form a train. A force is required to couple rail cars together. The way trains are assembled is to allow cars to roll downhill by gravity until they impact one another, and the impact force couples the cars. Unfortunately, this force is often excessive. Shell-and-tube exchangers are not designed to sustain this impact shock, which is transferred to the tube-to-tubesheet joints. These joints often fail because of it. Air coolers are much less susceptible to this type of damage, because it is easier to fit them with shipping pins designed to take the shock load. The pins are removed before the equipment is placed in use, a feature not easily built into shell-and-tube units.

21.2 ACCESS TO SITE

Once an exchanger arrives at a receiving dock or site, it eventually is moved to the installation location. Access is rarely a problem if the exchanger is being installed when the plant is being constructed. It is often a problem, however, if the exchanger is being installed as part of an upgrade of facilities, because other equipment may have been positioned in what had been the access path. This is particularly true in buildings or mechanical rooms containing HVAC systems. It is not as common in process applications, as these are mainly located out of doors.

21.3 INSTALLATION LOCATION

An air-cooled exchanger was selected to perform a cooling function. It was installed on a roof atop the sixth floor. To place it in position a crane was used. The crane rental cost and the cost of the exchanger were roughly the same. A slightly modified version of this exchanger could have been shipped knocked down, assuming no difficulties are encountered along the access path. The exchanger could have been reassembled on the roof for less than 20% of the cost of renting the crane.

Conclusion: Define the installation site and access path in the specifications.

21.4 MANUFACTURING LOCATION

Sometimes the handling or shipping limitation is a function of the point of manufacture. If a shell-and-tube exchanger is heavy or long, consider assembling it on a railroad car. This simplifies handling and could result in a reduced cost for both manufacturer and user. Further, there are areas, particularly in the south and southwesteren United States, where wider loads can be shipped by truck than in other areas. If both the manufacturing plant and installation site are in one of these areas, this manufacturer has a competitive advantage because they can offer larger, wider and more competitive units. A manufacturer's location can be particularly attractive in some situations if it is on or near water transportation. Shipping by barge offers the advantages of low cost transportation including the opportunity to select larger and more economical units. Make sure that winter conditions are factored in as waterways are not necessarily open the year round.

The potential for cost savings could be similar to the following. For large exchangers, the cost of shipping and handling is usually between 10 and 15% of the cost of the exchanger. If a suggestion given can be used, the savings in shipping costs might be in the 5 to 7% range. The larger units that could be offered due to shipping considerations might result in an additional 3% reduction in cost of the exchanger. Hence, an ideal exchanger selection will not necessarily be the most economical one unless the shipping method and charges are factored in.

Conclusion: All information that might be beneficial in achieving these savings should be in the specifications.

21.5 WIDTH CONSIDERATIONS

A 13-ft-diameter shell-and-tube exchanger was selected for an application. The installation site was roughly in the shape of a courtyard: access was between two buildings 12 ft apart. The way chosen for getting the exchanger to its final destination (it would not pass through this space) was to lift it over the building, adding to cost. Two smaller units in parallel would have allowed passage between the buildings and would have eliminated the need for a crane. However, two shells may not be a good choice in terms of plant operation. A larger tube diameter could be used, resulting in a smaller-diameter shell with longer-length tubes. (See Tables 2-1 and 2-2.)

The allowable shipping width when transport is over roads is not the same for all areas of the United States. Shipping width can be a limiting factor for one supplier and not a problem for another that has a more favorable manufacturing location relative to the job site. This often results in one or more suppliers' offering wider, longer, and more economical units than their competitors.

Conclusion: Define dimensional limitations in the specifications.

21.6 WEIGHT CONSIDERATIONS

Suppose the maximum capacity of a plant crane is 5 tons but the exchanger selected weighs 6 tons. Here again, an expensive crane rental could be avoided by using two smaller units. It is possible that tubes of longer length in a smaller-diameter shell would meet surface requirements and the weight limit as well.

Conclusion: Include known limits in the specifications. Also, see Chapter 13.

21.7 TRAVEL RESTRICTIONS

The majority of heat exchanger selections are not affected by travel restrictions. The limits usually apply to large exchangers that present problems related to shipping width, height, or weight. The engineer must decide with others, if the selection can be shipped without unreasonable charges, delays, or handling problems. These options are available.

Selecting another exchanger might be best with the popular choice being to reduce its size by using two or more smaller units. Power plant condensers are nearly always field assembled essentially avoiding the travel problem. Here, field assembly becomes the problem. A large unit could be shipped by rail using a dedicated train with no coupling of cars permitted. This avoids the shock load condition experienced when trains are assembled (Section 21.1). Sometimes, it is best to ship by barges which have large width and weight carrying capacities. Make sure a proposed water route will not be frozen if delivery is in the winter months. Shell-and-tube components can sometimes be shipped separately with the bundle, channel, header cover or lifting lugs shipped on one rail car and the balance on another. Air cooler columns can be designed in two pieces to reduce height to a manageable dimension. Another limit could be the load that the exchanger plus railcar impose on bridges over which they must pass.

The factors of length, width, height, and weight differ for rail, truck, barge, and air transport. Any one could be the determining factor. All must be taken into account before a recommendation is finalized.

21.8 FIELD ASSEMBLY

Before beginning the selection of an air cooler, information should be available about the quality and quantity of skilled personnel at the site to perform field assembly. If there are too few of them, it is probably best to ship factory-assembled units rather than relying on unskilled personnel. This and shipping needs should be addressed as one problem for air coolers (see Section 4.3).

21.9 MOUNTING PADS OR FOOTINGS

Air-cooled exchangers are provided with column base plates that rest on concrete pads furnished by the user. Manufacturers advertise that their units can be added to for future expansion. Side-by-side modules are predrilled to accommodate a future adjoining bay. Adding modules is a way of expanding capacity. The problem is that columns supporting a common bay carry about twice the dead load of the end columns on outside bays. To avoid future trouble during a plant expansion, which in effect increases the dead load on the end columns, make certain that concrete settings are the same for all. When end pads are too small, new end bays cannot be added easily without shutting down, designing new pads, and allowing concrete to settle. Making the pads alike avoids this costly and unnecessary problem.

21.10 A RADIATION CONDITION

For a discussion of this subject, see Sections 25.2 and 26.4.

Chapter 22

Relative Humidity

Heat exchanger engineers are sometimes asked what can be done to control the relative humidity (RH) in rooms where air is known to be too moist or too dry. In these situations a rule of thumb has been developed that can be beneficial in defining the problem and when quick answers are needed. The rule was developed using the known data that follows.

A/C units are designed for room air to pass through the evaporator and be cooled 20°F give or take a few degrees. Consider this example. The saturated pressure of steam in atmospheric air at 80°F is 0.50744 psia. Saturated pressures at other temperatures are given in the following table:

Temperature °F	psia
80	0.50744
60	0.25639
70	0.36334
50	0.17813
60	0.25639
40	0.12173

Source: Data tables in this chapter excerpted from *Steam Tables for Industrial Use*, American Society of Mechanical Engineers, New York, 2000, Table U-1, pp 169-71.

If this air/steam mixture is cooled to 60°F (20°F drop), its saturated pressure from the Table is 0.25639 psia. If it was exactly 50% RH, its pressure would be 0.50744/2 or 0.25372 which approximates the change in pressure that results from a 20°F drop in the ambient temperature. By comparing similar values at other temperatures, a rule of thumb can be developed. Air that leaves an evaporator saturated and that has been cooled 20°F will be delivered to the room at near 50% RH. The rule applies whether or not room air is saturated to begin with.

Consider a similar condition except that room air is cooled 16°F when crossing the evaporator. The following table results.

Temperature °F	psia
80	0.50744
64	0.29529
70	0.36334
54	0.20646
60	0.25639
40	0.14205
50	0.17813
34	0.09607

At 80°F the partial pressure of the steam is 0.50744 psia. When cooled to 64°F, its pressure from the steam tables is 0.29529 psia. The estimated 60% RH condition would be 0.6 (0.507440) or 0.30446 psia. These values are relatively close, never off by 1°F, so that a similar rule applies. If air leaves the coil saturated and is heated 16°F by room air, the room air will have a RH of near 60%.

These conditions occur frequently in air-conditioning systems. When saturated air leaves an evaporator the room RH depends on the temperature rise in the room. Approximations for other conditions are:

Cool air 12 *F	70% RH
Cool 8 air *F	80% RH
Cool 4 air *F	90% RH

This leads to several conclusions relative to the operation of A/C systems.

1. If there is a substantial reduction in an A/C unit's cooling load (examples: spring and fall season cooling loads are less, electronics may be turned off, summer days are cooler than design conditions), the A/C unit will be off line part of the time. When this happens room air will not be recirculated and its moisture will not be removed at the evaporator. This may result in a high room RH. If moisture must be removed, a dehumidifier or other means will be needed.
2. The A/C system alone cannot control RH over all operating conditions.
3. Suppose 40°F saturated outside air enters a building. If it is heated to 60°F, its RH will be near 50%. If heated to 80°F, its RH will be nearer 25%. (every 20°F change will double or half the RH depending on whether the air is heated or cooled.). A 25% RH is low and it will be lower if the air enters below 40°F. To increase the RH a humidifier will be needed.

In these examples the coldest temperature was assumed to be at the evaporator. This is not always true. Examples of this are noted in the two accounts of trouble jobs that make up the rest of this chapter. In one, the auditorium problem, the cooling load is small and moisture will not be removed because the A/C system will not come "on".

22.1 PARKED TRAILER OPERATING CONDITION

Assume a crew works in a parked trailer containing electronics racks. The air-conditioning or ECS system maintains an internal temperature of 70°F during working hours. At dusk the equipment is shut down for the night and the trailer sealed. During the night the ambient drops to 30°F and everything in the trailer approaches this temperature. Equipment and walls drop to a temperature that is less than that at the evaporator when it was turned off. When this happens moisture in the trailer condenses on walls, floor, electronics, and furniture. The result is wet floors and equipment. More important, moisture affects the performance of electronics.

The problem is correctable by installing a sensor to activate a heater when the temperature drops below a set value. This will cause the air-conditioning system to come on, condense moisture on the evaporator, and drain to the outside. Another solution is to allow fresh air to enter the trailer. In any event the engineer should be aware of this operating condition when designing mobile A/C units.

22.2 AUDITORIUM

Auditorium air-conditioning systems have different requirements from most A/C units. Their operating conditions vary over a wider range of conditions, as illustrated by the following example. Assume an auditorium is 40 ft by 40 ft by 12 ft high with a maximum occupancy of 250 people. The design ambient is 95°F. Consider three sets of conditions:

Calculated design conditions:

Heat entering through walls	9,600 Btu/h
Heat entering through roof	12,800 Btu/h
Lighting	18,000 Btu/h
Equipment	4,000 Btu/h
People 250 × (400 Btu/h)/person	100,000 Btu/h
Makeup air (7.5 cfm/person, including humidity)	66,000 Btu/h
Design load	210,400 Btu/h or 17.53 tons A/C

Consider the relative values in the calculated design condition which show that 79% of the cooling needed is people load and their fresh air needs. When this is a large portion of the A/C unit's size, questions similar to these should be asked. Does the maximum occupancy occur on a regular basis? Over what time period is the auditorium in use? Is it four or five times a summer for an hour or two at a time? Is it a one-time event, such as a Christmas party or a February stockholders meeting? None of these short term events would normally be the governing condition. Is the auditorium used more often but by fewer people? For example, if the number of people were 60 instead of 250, the calculated 17.3 tons of cooling could be reduced to 7.0 tons. The author cites this last example for a reason. Four or five times over a ten year period the author has seen A/C units selected for the maximum cooling expected to be encountered. In all of these cases, the results were oversized units that were off line for long periods of time. When an A/C unit is "off", the moisture in room air will not be removed resulting in rooms with high humidity, cold temperatures and much complaining about comfort. The problem is compounded by selecting the next larger standard size unit that meets the design conditions. The lesson is to refine the selection of the A/C unit using realistic scenarios.

Alternate operating condition 1. Assume the same auditorium, unoccupied, with lights and equipment off:

Heat entering through walls	9,600 Btu/h
Heat entering through ceiling	12,800 Btu/h
Lighting (turned off)	0 Btu/h
Equipment (turned off)	0 Btu/h
People (unoccupied)	0 Btu/h
Makeup air (assume 10% of max)	6,600 Btu/h
	29,000 Btu/h or 2.42 tons

Alternate operating condition 2. Consider the same auditorium on a cloudy, rainy fall or spring day when the ambient is 70°F and the relative humidity is 100%:

Heat entering through walls	0 Btu/h
Heat entering through roof	0 Btu/h
Lighting (turned off)	0 Btu/h
Equipment (turned off)	0 Btu/h
People (unoccupied)	0 Btu/h
Makeup air (assume 10% of max)	6,600 Btu/h*
Total cooling load	6,600 Btu/h or 0.55 tons

*In winter this value can be negative.

These figures illustrate the range of conditions (from negative to 17.53 tons) over which the A/C system must operate.

In general, A/C system humidity cannot be controlled over this range. The importance of each value should be weighed before design requirements are finalized. Here are things to consider:

- Oversized cooling systems compound the problem of humidity control.
- Consider alternate operating condition 2. The entering airflow might be for 25 people at 15 cfm, or 375 cfm, while the design load is 250 people

at 7.5 cfm (short-term occupancy), or 1875 cfm total. Note that the air changes per hour may be too low:

$$\text{Volume} = 40 \times 40 \times 12 = 19{,}200 \text{ ft}^3$$
$$\text{Makeup air} = 375 \text{ ft}^3/\text{mm} = 1 \text{ change every } 51 \text{ min}$$

By contrast, two or three air changes per hour are recommended.

When the cooling required is minimal, A/C units designed for a wide operating range will prove to be oversized. They will be turned on a minimum amount of time—not a sufficient amount of time to control room moisture. The result will be rooms with high relative humidities. These can result in irritating side effects such as

1. Pictures on walls wrinkling
2. Wallpaper staining.
3. Printing papers sticking together, wrinkling, or jamming the printing press
4. Uncomfortable personnel

The lesson is, Do not oversize an A/C system or you are apt to have a high relative humidity problem. When auditoriums operate over this range it is a good idea to use separate small and large A/C units to better control the extreme conditions. Small units will be "on" longer to remove air moisture.

PART 2

AIR-CONDITIONING EXCHANGERS

Chapter 23

Characteristics, Components, and Performance of A/C System Exchangers

This book has three parts, on process heat exchangers, air-conditioning heat exchangers, and cooling of electronics. It should not be assumed that exchangers can be conveniently separated into these categories. To illustrate, a short list of subjects from the process section (Part 1) that apply to A/C systems is given in Table 23-1, which lists sections and topics through Chapter 14 to illustrate the point. Only frequently occurring subjects are included in the table. Others that were passed over may still apply in some situations. Subjects are grouped in the table so that common problems can be readily traced.

The purpose of Part 2 is to alert engineers to problems that frequently prevent A/C systems from performing. For example, some exchangers might not meet some operating criterion thus making another selection preferable. The examples are intended to help identify problems that cause underperformance or designs that reduce system and other costs.

Heat exchangers in power plants, chemical plants, and refineries usually cannot be repaired unless the systems are shut down and given time to cool. This is not true of A/C system components or of some electronic problems. Many can be repaired while much of the system remains running. This can be a disadvantage. Personnel unfamiliar with A/C system details can make changes that reduce a system's performance without their or the engineer's knowledge. Often these are not documented. Different rules apply when trying to determine why A/C systems are not performing.

A/C exchangers can be rapidly selected because the properties of both streams are well known and thus simplified rating curves for their selection can be developed. Probably 9 of 10 exchanger problems that A/C manufacturers face are related to price, weight, and delivery. On the other hand, about 9 of 10 problems *users* of A/C equipment face are related to upgrading equipment, determining why a unit is not performing, addressing unsatisfactory air filtering, avoiding or correcting freeze problems, modifying systems to increase the quantity of makeup air, and matters related to entire systems more than to individual exchangers. This chapter is intended to assist in solving these problems.

Table 23-1 Some Topics from Part 1, "Process Exchangers," That Also Apply to A/C Systems

Chapter or section	Title or topic
Section 1.4.4	"An Underperforming Environmental Control System"
Section 1.4.10	"Noise"
Section 1.4.14	"Critical Pressure Drop"
Section 1.4.16	"Refrigerant Storage"
Section 1.8	"Cleaning and Fouling"
Section 1.9	"Contract Requirements between Buyer and Seller"
Section 2.3	"Parallel Flow"
Section 2.4	"Tube-Side Considerations in Selecting an Exchanger"; brazed construction; manifold header construction
Section 4.2.5	"Tube-Side Passes"
Section 4.3	"Structural Considerations" (factors that affect the choice of)
Section 4.6	"Plate Fin and Spiral Fin Extended Surface Exchangers"
Section 4.9	"Calculating Surface and Airflow Area for Fin Coils"
Chapter 5	"Extended Surface Metallurgy"
Section 5.1	"Selecting Aluminum or Copper Fins"
Chapter 6	"Air Cooler Fans or Blowers"
Section 6.1	"Fan Coverage and Bundle Geometry"
Section 6.1.3	"Sections Installed in Ducts" (duct coolers)
Section 6.2	"Fans and Blowers"
Chapter 8	"Plate Exchangers"
Chapter 12	"Freezing"
Section 12.1	"Antifreeze"
Section 12.3	"Heating Air or Water"
Section 12.4	"Air Recirculation"
Section 12.5	"Electric Heaters"
Section 12.10	"Other Reasons Why Coils Fail"; these are uncommon in process exchangers
Chapter 13	"The Iterative Process of Exchanger Selection"
Section 14.3	"Internally and Externally finned Tubes"

23.1 ADDRESSING PROBLEMS OF A/C SYSTEMS AND PERFORMANCE

Engineers are frequently asked to determine why an A/C system is not performing. To answer this, a place to start is the list of subjects in Table 23-2. Field trips are often necessary to identify problems. Some that frequently occur are the subject of later chapters in Part 2. In case studies, it is assumed that selections were thermally correct when installed. Factors that influence a selection are addressed in Part 2, as they were in Part 1 in connection with process exchangers. Most of the problems are described fully in other chapters and sections, but are listed in Table 23-2 so that in an emergency possible causes of the failure can be easily identified.

Table 23-2 Some Causes of A/C Systems Not Meeting Performance

Condensation
Drains, uncleanable A/C (Section 27.1)
Screen room cooling (Section 27.4)
Auditorium, alternate operating condition (Section 22.2)

Controls
Other reasons why coils fail (Section 12.7)
Air cooler fans or blowers (details of) (Chapter 6)

Ducting
Airflow in A/C and ECS systems (rule of thumb variable) (Section 19.4)
Louvered door replaced with solid door (Section 24.7)
Add a precooler in the makeup air line (freezing) (Section 16.3)
Screen room airflow insufficient (Sections 1.4.4, 19.3 and 19.4)
Takeoffs added (Section 24.8)

Heat exchanger (condenser)
Fan coverage insufficient (Section 6.1)
Noise (Sections 4.2.3, 4.3, and 17.3)
Operating pressure not within design limits (solar radiant energy, Sections 18.11 and 21.10)
Recirculation (Chapter 20)
Starving of fans (Section 20.2)
Radiation, solar (Section 18.11 and 21.10)

Heat exchanger (evaporator)
Freezing (Chapter 12; Sections 20.4 and 23.3)
Noise (Sections 4.2 and 17.3)
Operation at a condition other than design (Section 25.1)
Replacement of evaporator on basis of dimensions only (Section 25.4)
Replacement of evaporator with one having different diameter tubes (Sections 2.4 and 25.4)
Superheating (Section 23.3)
Undersized heat exchanger (Section 23.3 and Table 29.5)

Other operating conditions
Drains (freezing or staining) (Sections 12.2 and 27.1)
Electric heaters not on when needed (Section 12.5)
Freezing (Chapter 12; Sections 12.2 and 12.7)
High pressure (Sections 18.11 and 21.10)
Insulation material (Section 24.1)
Mixing insufficient or too late to avoid freezing (Section 20.4 and 23.3.1)
Noise
  Condenser Fans (Section 17.3)
  Ducting (Sections 17.1 and 17.3)
Evaporator (Section 17.1)

(continues)

Table 23-2 ***(Continued)***

Performance
A/C not running full time, "on" cycle too short (Section 25.1.2)
Air precooler needed to cool makeup air (Section 26.1)
Alternate operating condition (Section 22.0)
Blockage
  A/C duct changes may reduce airflow (Chapter 24)
  Air filters upgraded or added in series (Section 24.5)
  Duct continuity break (Section 24.3)
  Duct geometry limits flow (Section 19.3, 19.12, Chapter 30, Figure 19-2 and Table 29-1)
  Duct linings (Section 24.1)
  Duct resistance increased (Sections 24.5 and 29.2)
  Duct size (Section 29.2)
  Duct turning vanes or lack of (Section 24.4)
  Ducts changed from their original design (Section 24.2)
  Equipment placed in front of return duct (Section 24.6 and Figure 19-1, example 4)
  Equipment positioned on top of return (Section 24.6)
  Equipment mounted too close to building (Section 20.2)
  Field modifications (Chapter 23)
  Turning vanes didn't hold position (Section 24.4)
Chiller or ECS operated at a lower than design evaporating temperature (Sections 25.1and 25.3)
Cost
  Refrigerant quantity (Section 15.1)
  Reduced horsepower (designs to meet performance at reduced energy cost) (Sections 14.8 and 17.5)
Cooling load increased/decreased (Chapters 25 and 26)
  Condensation (Chapter 22)
Evaporator too small (Section 23.3)
Evaporator air velocity above 500 ft/min (Section 17.1)
Field modifications changed system design (Sections 24.1, 24.2, 24.5 and 24.8)
Filters (Section 24.5)
Insulation added to building (system changed to meet performance) (Section 26.2)
Makeup air quantity—and therefore cooling load—increased (Sections 24.9 and 26.1)
Mixing inadequate (Section 20.4)
Noise (Sections 4.2 and 17.3)
Oil return by gravity inhibited (Sections 2.4, 23.3, and 23.4)
Overcharged system or one low on refrigerant (Section 25.1)
Radiation
  Cable (Section 18.12)
  Circuit breaker tripping (Section 18.12)
  Condenser operating temperature increased (Sections 18.11 and 21.10)
Recirculation (Chapter 20)
Resistance increased or duct blocked (See Blockage)

Relative humidity (Chapter 22)
A/C not running a significant length of time (Section 22.2)
Reduced cooling load (Section 22.2)
  Summer (Section 22.2)
  Winter (Section 22.2)

23.2 A/C SYSTEM COMPONENT EXCHANGERS: EVAPORATORS AND CONDENSERS

The main exchangers in A/C systems are the evaporator and the condenser. These can be air- or water-cooled. In a typical air cooled installation the surfaces of both exchangers are manufactured of the same material and thickness of fin (aluminum) and tube (copper base). One manufacturer is usually the supplier of both exchangers. This is not a requirement, but the manufacturer of one exchanger will have the die needed for punching the fins to match. Dies are expensive. Where the construction of these exchangers differs is at their headers.

Each exchanger's volume is usually less than 5.6 ft^3, so most do not require a code stamp. Tube ends are sealed by brazing, at least on small units. Brazed seals are visible from the outside, making it easier to find leaks than in shell-and-tube exchangers. The tubes are often of copper base materials, for cleanliness and to avoid contaminating the refrigerant. Tubes are seldom thicker than 18 BWG. Aluminum is the common fin material because of its good thermal properties and low cost. Remember, in A/C applications, tubes are measured on the inside diameter. As an example a 5/8-in A/C tube is 0.625 inch in inside diameter, while in process exchangers a 5/8-in tube is 0.625 inch in outside diameter.

The typical plate-fin tube will be thin and of materials with low tensile stress. These are easily expanded to create contact between tube and fin. The tube ID will increase about 3% in the expansion process. Many U-bend sizes are available, as a large market for them exists. These make available a large number of pass choices on the tube side (Section 2.4).

The manufacture and assembly of short-length tubes is rapid using plate fins, compared with the complexity of winding spiral fins on tubes. Fin unraveling is not a problem. The fin height can be greater using stamped fins because tube holes are punched to suit, provided the proper size die is available.

Spiral fin heights are limited by the amount of stretching that aluminum or copper or other materials can sustain. Spiral fins are thinner at their tip than at their base, while plate fins are of near constant thickness. Spiral-fin tube spacing depends on the tube support designs available.

Evaporators and condensers differ in specific ways. Evaporators are mounted indoors and do not experience the air temperature swings of condensers. Evaporators sometimes contend with icing, while condensers will almost always melt any ice present. This is usually not a problem in their operation. The refrigerant that enters a condenser is a vapor under pressure that can at times be operating below atmospheric pressure. The refrigerant is distributed to many tubes in parallel. Condenser headers must have adequate cross-sectional area. On the other hand, vapor-liquid mixtures enter evaporators following passage through a distributor and thermostatic expansion valve. The purpose of the distributor is to avoid the possibility of one evaporator circuit, by chance or accident, receiving a greater share of liquid than another. If this happens it would make the evaporator inefficient. Unbalanced flow can occur when thermostatic expansion valves only are used.

Evaporators are designed to mate with duct systems. They are positioned between transition pieces so that air will be evenly distributed across the coil. They are matched with filters and any heater required. Once they are in position the system is difficult to modify without time delays and high cost. Evaporators should be designed to avoid freezing. They stand alone and are independent of heaters, ductwork, and air filters. When necessary, they can be replaced or increased in size, but this is seldom done at low cost.

23.3 AIR-CONDITIONING SYSTEM EVAPORATORS

The following are requirements that apply to the sizing of evaporators.

The air velocity through the evaporator is limited to 500 ft/min measured at the gross face area of the coil. At higher velocities, air will blow condensate off the coil (see Section 17.1) and into ductwork, causing rusting or other problems. An exception is environmental control system (ECS) evaporators. In this service, the makeup air is often taken from the building rather than from the outside. Its moisture content is low because it was removed by the building A/C system. Other information on this subject is given in Section 19.4.

The refrigerant that leaves an evaporator should be superheated, usually 6 to 10°F. The superheating section will vaporize nearly all the liquid present, thus keeping liquid from entering the compressor. Compressors are designed for vapor or gas service, not liquids. Compression of liquids may damage compressors. Superheating vapor minimizes liquid carryover. If superheating is too high, the compressor's cooling capacity will drop.

An essential property of evaporators systems is that no chamber or pocket should exist that allows compressor oil to accumulate and be trapped as it circulates through the system. Tubes should be positioned horizontally, as this will allow oil to return to the compressor by gravity. This applies to the complete system including piping and condenser. The oil can be raised to a higher elevation using a double-pipe riser. This subject is addressed in many A/C books.

The pressure drop in an evaporator should be no greater than the difference between the entering evaporator pressure and the pressure that would exist if the evaporating temperature were 2°F lower. This means that a refrigerant in a system operating at 100 psia will have a higher allowable pressure drop than that in one operating at 15 psia.

There is little that can be done to increase the capacity of undersized evaporators. In some cases the capacity can be increased up to 5% by increasing the quantity of air flowing. This is not always possible, as the 500 ft/min velocity limit must be met. A higher-horsepower blower motor may be needed to make this change. Noise and space present other limits. When in doubt, select a larger evaporator, as there are ways of reducing its capacity. Never size an evaporator using curves from unknown sources. Published curves are not always accurate, and evaporators have been undersized for this reason. Always use data known to be reliable when sizing a unit.

Evaporators and condensers should be selected with small-diameter tubes to minimize the quantity of refrigerant required and thus to save on cost (see Section 15.1). The choice may result in a small increase in the cost of the evaporator but will reduce the overall cost of the system.

Feeding refrigerant to an evaporator differs from flow into other exchangers. In most exchangers a minimum pressure drop at the inlet is advantageous. The evaporator pressure will be lower than in the liquid line that feeds it. A large pressure drop occurs in the thermostatic expansion valve (TXV), distributor, evaporator inlet, and tubes. The idea is to have evaporation take place (in the evaporator) at a lower temperature (translation: lower pressure) and have condensation occur (in the condenser) at a high temperature (translation: higher pressure). The difference in these pressures is roughly equal to the difference between condensing and evaporating pressures. The distributor when properly designed has small-diameter tubes in parallel, which provide the high pressure drop needed. These tubes are designed to distribute equal quantities of a liquid-vapor mixture to tubes in parallel so that the refrigerant does not bypass surface. The TXV and distributor control the operating pressure. This is not possible if the evaporator is undersized. In an undersized evaporator not all refrigerant will vaporize, and therefore the evaporator pressure will rise. The result is a rise in the room temperature and a decrease in the quantity of available cooling.

Conclusion: Size evaporators correctly.

23.3.1 Some A/C Evaporators That Did Not Perform in Service

Three case histories of evaporators that did not meet performance are cited in this subsection. In each case the cause was beyond the control of the manufacturer. Items 1 and 2 were caused by failing to account for alternate operating conditions in the specifications, while case 3 is the result of a poor duct design. The problems were:

1. An oversized system
2. An icing condition (tube-wall temperature)
3. A duct design that causes evaporator icing

Oversized System

Case 1 is a relative humidity problem due to an alternate operating condition. It was discussed in Section 22.2.

An Icing Condition—Tube-Wall Temperature

In case 2, the refrigerant evaporating temperature decreased as the ambient decreased. The ambient can reach 20°F without formation of ice on fins. The reason is that metal temperature determines whether ice will be present, not ambient temperature. The possibility of lower ambients should be in the specifications. The temperature drop across tube, fin, and fouling layers can prevent ice from forming even though the temperature in the tubes is near 32°F. Icing on the external surface of evaporators can be eliminated using temperature-controlled heater(s) designed to raise the ambient above the freezing point. Note that the metal temperature will change rapidly if units are changed from operating to shutdown conditions. This is a normal operating sequence. The metal temperature may be at or above freezing during operation and below freezing following shutdown. This condition can be the cause of freezing on either side of tubes.

A Duct Design That Can Caused Evaporator Icing

A given amount of cold outside makeup air was to mix with building recirculating air before passing over a coil. In theory the mixing of makeup and recirculating air would prevent freezing. However, freezing occurred because insufficient space was provided for complete mixing of the two streams. The necessary condition is that the combined stream experience one turn and travel seven equivalent tube diameters for mixing to be complete. Two kinds of freezing can occur. If the coolant is water, it can freeze inside the tubes. If the tube metal temperature is below freezing and the unit is shut down, moisture can freeze on the outside of tube and fin. (For a discussion of this see Section 20.4). Either condition can cause the evaporator to fail but freezing in a tube or tubes causes failure more often. When many plate-fin tubes are cracked, the coil must be replaced, which is costly. If few are cracked, they can be plugged. Individual tubes in spiral-fin units can be replaced. Replacing cracked tubes does not solve the problem that caused the failure. Corrective steps must be taken. After the building's cooling system is complete, modifying the installed ductwork is usually not easy. Different installation problems require different solutions, but one thing is certain: costs and delays result.

23.3.2 Evaporator Superheating Zone

Rapid boiling in evaporating and vaporizing processes often carries drops of liquid along with vapor. The problem is that in A/C systems this liquid will go directly to the compressor. Compressors are designed to compress gases or vapors, not liquids. Liquids can damage compressors. For system integrity and to avoid liquid carryover, A/C evaporators are designed with a superheating zone to assure liquids are vaporized. Typically, superheating is in the 8 to 10°F range, but rarely less than 4°F. In sizing this zone remember that evaporating and vaporizing rates are high while rates for superheating gases or vapors are low. This differs from steam operating in the desuperheating zone. In this application high LMTDs occur when heat transfer rates are low, and relatively low LMTDs occur when boiling rates are high. In A/C evaporators, both evaporating and superheating zone LMTDs are low. An undersized desupercheating zone does not cause physical damage to the evaporator but the liquid carry over that results is often the cause of compressor failure.

23.3.3 Undersized Evaporators

In this book all "problem" exchangers described are assumed to be thermally correct except for undersized A/C evaporators. Evaporators are frequently selected from curves that combine refrigerant rate with metal and fouling resistances. Unfortunately, not all published curves are correct, and so this selection method has resulted in undersized evaporators. When an evaporator is too small, the problem is difficult to correct, because space is needed and is seldom available to install a larger evaporator. The result is that if the needed space for the evaporator is not available problems similar to the following must be faced:

1. Larger duct transition pieces are needed to mate with the larger evaporator.
2. The transition pieces must be lengthened for good air distribution.
3. Heaters and filters must be modified to match the face area of the new coil.

Additional surface could be obtained by adding another layer of tubes. This may result in a simple fix that requires removing only a short length of duct to accommodate the added row. The removal is not always possible as it depends almost totally on whether space is available to make this change.

The best solution is to select the evaporator correctly before the system is built. The downside in doing so is that undersized evaporators usually cost less, which may be why they are sometimes chosen. Fortunately, undersized evaporator problems are becoming less common. When they occur, they are difficult to correct due to limiting system dimensions.

Recommendation: Available tube and fin sizes and spacings depend on die sizes a manufacturer has on hand. They are not necessarily the same as those of competitors. For a given application most air-cooled evaporators, regardless of tube size and spacing, occupy about the same volume. Keep a known-to-be-accurate curve on hand to compare offerings. If the recommended evaporator's volume is less than 85% of this known curve's offering, the chances are it is undersized (see the opening paragraphs of Section 23.3).

The following situation suggests an evaporator (or an A/C system) that is undersized. An area served by the evaporator is designed to operate at 75°F and may be cooled only to 83°F at full-load conditions. At this temperature the discomfort level of people working in the area is a problem needing correction. Begin by checking the evaporator size using the information from the known curve as a reference.

The above example is one way the problem surfaces. Another is when electronics fail or become overheated. The cause may be difficult to isolate for the following reasons:

1. It may be an electronics system problem outside your domain.
2. A/C or ECS units are furnished with evaporators selected by the manufacturer. Rating data for the specific evaporator in service are usually not available when corrective action must be taken. This is when the reference curve is beneficial.

23.4 A/C SYSTEM CONDENSERS

Evaporators and condensers in a given system are usually constructed with the same tube size, spacing, and fin type, although it is not necessary that their surfaces have the same geometry. Condensers are similar to evaporators in the following ways:

1. Each is sized using the same fouling factor on the refrigerant side.
2. Each has noise limits.
3. Refrigerant migration must be addressed for the system.
4. Each must allow all oil to return to the compressor.
5. Both units can be selected to minimize the amount of refrigerant in the system.
6. Neither performs when the refrigerant charge is lost.

Condensers differ from evaporators as follows:

1. The condenser inlet pressure drop should be minimal. In the evaporator the pressure drop across the thermostatic expansion valve, distributor, and evaporator inlet control the internal pressure. This combined pressure drop is large.
2. Condensers are often purchased with propeller fans. The evaporator air mover is often a blower that provides the pressure needed to move air through the evaporator and the duct system, across air filters, and past heaters.
3. Condensers usually do not need heating coils to prevent freezing.
4. Condensers are not limited by coil aspect ratios.
5. Condensers are often standardized and may be oversized for an application.
6. Condensers do not have a separate condensate heat load as evaporators do. Relative humidity is not a factor in their sizing.
7. Condensers are not limited by a 500 ft/min air velocity across the coil.
8. Condensers are usually mounted out-of-doors.
9. Inadvertent blocking of airflow is rarely an evaporator problem. Locating a condenser too close to a building, wall, or heat source sometimes causes a recirculation or starving the fan problem.
10. High head pressure should be addressed particularly if the condenser is subject to solar radiation (Sections 18.11, 21.10, 25.2, and 26.4).
11. Condensers are usually purchased with a propeller fan(s) as part of the system. Evaporator blowers are not normally purchased as part of the evaporator.
12. The condenser operates over a greater range of temperatures than the evaporator.

Chapter 24

Underperformance in A/C Systems Due to Ducting Problems

Consider three rooms, each the same size and construction, and each cooled in a different manner:

- Room 1 is cooled by cold air supplied through a duct.
- Room 2 is cooled using a direct-expansion A/C system.
- Room 3 is cooled using an air-handling unit and cold water from a chiller.

Now assume each room's cooling load is doubled as a result of a similar change occurring in all rooms. In each case the exchanger surface for cooling the room and the required flow will be doubled. Supply lines become the problem. For the direct-expansion and water-cooled systems, it is relatively easy to make needed changes by increasing the line size from $^1/_2$ to $^3/_4$ in or from $^3/_4$ to 1 in. Space is almost always available to make changes of this type. However, it is rarely available when the A/C method is air cooling, because the original ducts are large in size and simply cannot be increased by 50 to 100% in area because larger ducts will likely interfere with existing ducts.

Thus, it is more difficult to supply added cooling to an area that is cooled by air than by a water-cooled or direct-expansion unit. In air-cooled units, a problem *generally* occurs when the room's cooling load is increased substantially, requiring more airflow. Flow can rarely be increased by more than 15% without changing the supply duct. The solution is not easy, on account of space constraints, and must be found on a case-by-case basis. If it is thought that a room's cooling load will increase shortly after the system is installed, another cooling method should be considered, or ducts should be sized to accommodate the future need. When determined to rely on an air-cooled unit, always size and locate ducts so that the system can be modified to increase capacity. This is not an easy task.

A way of accommodating an increased cooling load when using an air-cooled unit is to leave the duct system as is and provide a water-cooled system *locally*. Substantially smaller line sizes are needed to deliver cool water to a water-cooled exchanger or to a direct-expansion unit than to deliver air. The add-on equipment, of either type, should be located in or near the room being cooled. If the increased load includes people, this recommendation may not do, because it does not provide for an increase in makeup (i.e., *fresh*) air.

Heat exchangers installed in ducts will generally, but not always, fail to meet performance if the duct system is undersized. The rest of this chapter will describe several other causes of underperformance, often citing examples. In all "problem" installations described, heat exchangers and duct systems are assumed to be thermally correct when selected, except where noted. It is nearly always easier to make changes to ducts than to heat exchangers, and thus underperformance sometimes turns out to be the result of unauthorized field changes, which can often be made without interrupting service. Most inside duct conditions are not visible from outside, and thus it is often the case that the cause of a problem is elementary but locating and correcting it is not.

24.1 SOUND-DEADENING OR INSULATION MATERIAL ADDED

A 20- × 12-in duct was selected. In operation, it was found that the air noise level was high. To reduce it, 1 inch of noise-deadening material was added. After the upgrade, the inside duct dimensions were 18 in × 10 in. The smaller airflow area, however, caused an increase in air *velocity,* itself a cause of increased noise. A similar reduction in airflow would exist if the same thickness of insulation had been added internally. Both conditions increase the system pressure loss and can result in a reduced cfm supplied by the blower, which can be the cause of a drop in cooling capacity.

It is obvious that adding noise-deadening material may not be a good solution. While in principle it dampens noise, it can simultaneously increase it by increasing the air velocity. The trade-off is to increase the size of the duct.

24.2 DUCTS NOT BUILT IN ACCORDANCE WITH CALCULATIONS

Examples of duct systems not built the way they were designed, thus preventing environmental control systems from performing, were given in Sections 1.4.4, 19.3, and 19.4. The lesson is, when checking systems for underperformance, make sure the balance of the system was built as designed. This is usually the user's responsibility and outside the control of the ECS manufacturer.

24.3 BREAKS IN DUCT CONTINUITY

The problem of breaks in duct continuity does not occur frequently. When it does occur, it is usually in older buildings. During modifications for new tenants, existing ducts are changed. Sections are removed to provide space for other needs. After modification one end may be left uncapped, delivering cold air to parts of a building where it is not wanted. Sometimes ducts are disconnected near the cold-air source but small booster fans, installed earlier in the disconnected duct, continue to run, so that air movement is still observed through the duct. Movement is there, but it is not air that has been cooled. This problem is difficult to locate because it is unexpected and almost always occurs in areas difficult to access. This is probably why the problem wasn't corrected earlier.

Conclusion: The lesson from such an unexpected cause of underperformance is, Do not assume anything.

24.4 DUCT TURNING VANE FAULTS

Poorly designed turning vanes installed in elbows can increase the turning loss many times. If turning vanes are not installed or built properly, the system pressure drop can increase significantly, resulting in reduced flow. To achieve minimum turbulence, vanes should be positioned so that turns are smooth. Instances are known in which poor-quality vanes were built and installed correctly but their built in locking device was not adequate to hold them in position. The blades rotated in service and partially blocked flow. The vanes did more harm than good. A reliable locking device is needed to hold vanes in position.

The built in locking device construction is a problem not often faced, and is difficult to identify because it is not visible from outside a duct.

24.5 AIRFLOW REDUCED BY AIR FILTERS

A discussion of air filters is in order. Filters that remove large particles from air streams generally have a low efficiency and pressure drop. This condition may have been part of the pressure loss used in the blower selection. Assume smaller particles must be removed requiring a more efficient filter. The more efficient filter needed will remove the particles but at a reduced air flow and increased pressure loss. If even smaller particles must be removed a more efficient filter is needed further reducing flow and increasing pressure drop. A possible modification is installing two filters in series. One is more efficient and costlier; the other less efficient and less expensive. The low-cost filter will remove the large particles and be replaced more often. In this way, the expensive filter is replaced less often saving on cost. The end result is that as filter efficiencies increase, pressure losses increase. Under these conditions the air that the fan or blower can deliver decreases.

Consider the effect this will have on the ECS or A/C system. Blowers provide air at the guaranteed pressure drop. If the system pressure loss rises, the air circulating will be reduced and the evaporator will no longer perform as less air flows across it. The trade-off is cleaner air and reduced cooling or less efficient filters and a more efficient cooling system.

Filters have been upgraded without considering the effect this has on cooling and without the engineer's knowledge. This can cause electronics to overheat. If a cooling unit is not meeting specifications, check if the filters were upgraded which can be easily done as they are accessible.

Conclusion: Before upgrading air filters determine the effect that replacements have on system performance.

24.6 BLOCKING OR ALTERING A SUPPLY OR RETURN

Section 19.3 recalled an ECS that did not perform because the ducting pressure drop was too high, above what the system was designed for. The limiting pressure loss was established by the heat transfer engineer. This limit was later altered by others, unaware that the changes they made altered the performance of the unit. One can engineer a system correctly and find that it does not perform in service. Engineers faced with determining why a system is not working should look for conditions much like those that follow.

Just as in the screen room example of Section 19.3, pressure drop can be added to the system in several ways. The most common is to block the air supply or return. It is unlikely the heat exchange engineer will be aware of these changes. Two examples are through starving a fan and through bypassing. Other examples include placing furniture or cabinets against a wall effectively blocking an air supply or return; and placing equipment on top of an air supply or return inlet or outlet. (See Figure 19-1, item 4.)

Here is a case that occurs more often than one would like. EMI/RFI filters are designed to keep stray electronic signals in a room or rack. When air passes through them a pressure drop results, of approximately 0.1 inHg. Most of this occurs in straightening of the airflow and minimizing of any eddies. If two filters are in series, the pressure loss will be only about 0.115 inHg, because all airflow straightening will have been done entering the first filter and is not repeated in the second.

Now consider a bank of racks where an existing 21-in rack is replaced with a 31-in rack. All other racks will be relocated a few inches to the right or left. This moves the floor openings, relative to each other, a few inches to the right or left. The in-line position of these openings no longer exists. The pressure loss across them will now be nearer 0.2 inHg or more. Seldom is this considered when racks are relocated. Depending on how much the filters are out of line, a turning loss may also have to be planned for. In addition to this, one should make sure that the floor opening remains beneath the rack. Often it is partially blocked by the adjoining rack which was moved.

24.7 LOUVERED DOOR REPLACED

In some A/C installations rooms and corridors serve as return air ducts. Doors in the return should be louvered to allow warm air to return to the evaporator even though they are closed. It occasionally happens that personnel unfamiliar with A/C requirements decide to upgrade these doors and replace them with solid ones. If you enter an area for the first time and a louvered door was replaced, it is unlikely you would know that one was once there. You can, however, look for an open air return path. If none is present it is a possible cause of underperformance.

24.8 TAKEOFFS ADDED REDUCING AIRFLOW TO OTHER ROOMS

There are times when the cooling needs in an area of a building increase because of a change in room usage: the number of people using the room increases and equipment is added. Both add to the local cooling load. To increase flow to these rooms, takeoff(s) are added in nearby duct(s), reducing the quantity of air delivered to other outlets in these lines. Reduced flow can cause the temperature to rise in the other rooms. In general, adding takeoffs results in undersized supply ducts. This can be corrected by increasing their size. Most changes in flow to a group of rooms will not be enough to warrant changing the duct size. Realistically, these ducts should instead be rebalanced to arrive at a workable solution.

24.9 MAKEUP AIR QUANTITY INCREASED

The makeup air duct may have been increased in size so that more fresh air is provided to a room or area. If this was done, it will have added to the cooling load on the A/C system. It should be determined if the increased cooling load resulting from more makeup air is the cause of underperformance.

Chapter 25

Other Conditions That Cause A/C Systems to Underperform

25.1 OPERATING CONDITIONS THAT REDUCE THE CAPACITY OF EVAPORATORS

Engineers are usually faced with three kinds of evaporator problems:

1. Sizing the evaporator initially
2. Determining why an evaporator of their own design may not be performing
3. Determining why an evaporator sized by others on which they have no data is not performing

The subject being addressed concerns points 2 and 3 only. Engineers are advised a unit has been in service several years when they are asked to determine why it is not performing. Here are potential causes of the problem.

25.1.1 System Operated at Reduced Evaporating Temperature

A/C manufacturers publish curves defining the performance of their equipment at various operating conditions. An example of this is shown in Table 19-1. It has been the author's experience that units purchased to perform at one set of conditions (nameplate conditions) are often operated at other conditions. Reviewing the data given in Table 19-1 gives an idea of how much cooling is gained or lost when this change is made. Nearly always, the cooling capacity at which a system is operated is less than the capacity for which it was purchased.

25.1.2 Cooling Load Reduced Substantially

A/C units are designed not only to cool air but to remove moisture as well. The latter is necessary to lower the relative humidity in rooms or areas. When a room's cooling load is low (see Section 22.2), the A/C unit can easily provide the cooling needed. However, if air in a room is to be changed three times per hour and the A/C unit is on only five minutes of every hour, the A/C will not be on long enough to remove a sufficient amount of moisture to make the room comfortable or provide enough fresh air. In this case most room air will not pass through the evaporator, where moisture is removed. Here's why:

- One air change takes place every 20 minutes if the unit keeps running.
- A $^{1}/_{4}$ air change takes place during the 5 minutes the unit is running.

- Three-fourths of the air in the room does not pass through the evaporator to have its moisture removed.
- Hence, under these conditions, a room with a high relative humidity will remain at a high relative humidity.

25.1.3 Refrigerant Charge

Air-conditioning units are designed to operate with a quantity of refrigerant that, during operation, maintains the pressure and temperature in both evaporator and condenser at a given set of conditions. The amount needed will vary depending on the size and length of interconnecting piping and the volume in the exchangers. Installations with primary and remote condensers that run at alternate operating conditions are seldom effective because the length of piping needed and hence the charge needed differs due to the different pipe lengths in each system. Further, when switching from one condenser to the other, it is difficult to know how much refrigerant will remain in the condenser and piping that is now idle. If the quantity of refrigerant in the operating system is low, the pressure in both evaporator and condenser declines and the system will not provide the desired cooling.

If the system charge decreases sufficiently the evaporator pressure will decrease and may drop to a temperature low enough to cause icing on the outside of the finned tube. If this happens, some airflow will be blocked thus reducing performance. On the other hand, if the charge is too high the pressure in both evaporator and condenser rises, requiring more energy to run the system. Summed up, a less than sufficient charge will not deliver the required cooling, while overcharging will increase the cost of operating the system.

25.2 CONDENSER SUBJECTED TO SOLAR RADIATION

Consider Genetron R 502. Its operating pressures and temperatures are:

Operating temperature, °F	Operating pressure, psia
95	216.1
100	230.9
105	246.4
110	262.6
115	279.6
120	297.4
125	316.1
130	335.5
135	355.9
140	377.3
145	399.6
150	423.1
155	447.6

Permission of Honeywell Refrigerant Technical Service.

Suppose a condenser has an operating temperature of 130°F, which corresponds to an operating pressure of 335.5 psia. If the condenser is located in the bright sunlight on a 120°F day, it will receive radiation from the sun that will raise the metal temperature roughly 20 to 25°F. This raises the tube temperature to 155°F (130 + 25). At this temperature the corresponding refrigerant pressure is 447.6 psia. Most systems are not designed to sustain this pressure. To protect against high pressure, a cutout switch is installed that will shut the system down when the pressure reaches its design limit. A/C systems are not always designed for the high pressures that can be reached due to solar radiation.

Recommendation: In positioning the condenser, avoid placing it where it receives direct sunlight so the problem can be avoided. Also, make certain the maximum operating pressure is within the system's pressure limit. A completely assembled system may need to be protected against solar radiation during shipment.

25.3 ECS SYSTEMS OPERATED AT OTHER THAN DESIGN CONDITIONS

An alternate operating condition for an ECS system is also an alternate operating condition for its evaporator.

Suppose an ECS system is to operate at 80°F DB (dry-bulb), 67°F WB (wet-bulb). The unit selected to supply the desired cooling is guaranteed to deliver 379,000 Btu/h; see Table 19-1. Now suppose the user operates the same ECS system at a different room temperature, 72°F DB, 60°F WB. In doing this the cooling capacity is reduced to 338,800 Btu/h, a drop of 10.6%. The refrigerant's evaporating load has been reduced and its heat of compression load increased. As in all exchangers the evaporator's cooling capacity changes because its operating condition has changed. Look for an alternate operating condition as a possible cause of underperformance of the system. Also, be alert for the same problem in chiller operation.

A variation of the above is a frequent cause of trouble in the performance of chillers. ECS units are selected with a high sensible heat ratio because much of their condensing load has been removed in a building's A/C system. When a chiller is selected it has a higher heat of condensation to be removed than do ECS unit(s), at least on a percentage basis. The sensible heat loads of these units differ. Often the total cooling supplied meets the total cooling needed but the load is not properly apportioned. It doesn't meet the application need for more sensible cooling. This is a frequent cause of poor chiller selection in buildings that have high electronics cooling loads and no screen or computer rooms. The problem is that insufficient sensible cooling is provided. The problem becomes known when rooms become hot. Look for this as a possible reason for underperformance.

25.4 REPLACING COILS SOLELY ON THE BASIS OF SIMILAR DIMENSIONS

Substitution of an incorrect coil commonly occurs when a home air-conditioning system evaporator or condenser must be replaced. The service technician, in an effort to complete the repairs, asks the user to approve the use of a coil that has approximately the same external dimensions and that is immediately available. Approval should not be granted without confirming the heat transfer and pressure drop values of the proposed unit. A change in tube diameter (the usual case) has a pronounced effect on the tube-side and the overall heat transfer rate. The length of travel increases with increasing tube diameter (see Section 2.4). Hence, replacing a coil with one having larger-diameter tubes should result in a coil that has more tube-side passes (rarely mentioned). An engineering rating is required to confirm that the proposed coil will meet the performance of the coil being replaced. Service personnel rarely provide this verification. If the substitution is allowed the result is often replacing a properly sized evaporator with an undersized one or one that will lead to other kinds of trouble.

Chapter 26

Ways of Increasing the Cooling Capacity of a System

Once an evaporator system (coil, heater, filters, louvers, instrumentation) is in place it is costly to modify the system in terms of dollars and downtime. As an alternative, consideration should be given to increasing the available cooling without making changes to the existing system. Here are ways this can be done:

1. Add a precooler in the makeup air line.
2. Insulate the building.
3. Add a water-cooled unit in the area to be cooled.
4. Install a sun shield above the condenser.
5. Spray the condenser with water.
6. Provide colder water to air-handling units.

26.1 ADD A PRECOOLER IN THE MAKEUP AIR LINE

The evaporator selected has been performing as designed. Later, a change is made increasing the cooling required. It could be that the electronics cooling load increased or more makeup air is needed. In either case cooling can be provided by enlarging the makeup air duct and installing a cooler in this line. Usually this unit is water-cooled. In effect, this is a remote evaporator installed to cool makeup air only. This change can usually provide sufficient cooling for makeup air. If the equipment load is increased substantially, it is doubtful this design change will meet the new requirements.

A word of warning if a remote evaporator is installed. The air that enters the main evaporator is at the mix temperature of room and outside air. The air that enters at the remote evaporator is at the outdoor air temperature. Draining the remote evaporator at times is a must to avoid freezing the coil. A heater could be added, but it must come on when needed to prevent freezing.

Adding a coil in the makeup air line is much less work and requires much less downtime than making modifications at the main evaporator and ducting. A new section of makeup air duct plus transition pieces, elbows, and coil is needed and can be on hand, assembled, and ready to install before any change is made to the existing system. In a nonpeak period, the A/C system can be turned off and an existing section of makeup duct can be replaced with the new coil and ducting. The newly assembled system will likely be back on-line in 8 hours or less. This may translate to no downtime. Changes at the main evaporator take substantially longer, roughly two weeks, because a larger coil is needed. This means modifications at the heater, filter, and transition pieces and providing space for their installation. The system is off-line while this work is done. Installing a new preassembled system at this location is usually hard to justify economically.

Cooling provided at a remote evaporator must be sufficient to satisfy the new requirement when added to the existing system. Otherwise, adding a coil does not solve the problem but merely minimizes it.

26.2 INSULATE THE BUILDING

It is possible to increase the cooling available by adding insulation to a building, thus decreasing its heat gain. The cooling saved can be used for cooling in other areas.

26.3 ADD A WATER-COOLED UNIT IN THE AREA TO BE COOLED

A water-cooled exchanger can be added to cool an area. The water may be from a chiller, well, or other source. This choice takes advantage of the conditions discussed in the introductory paragraphs of Chapter 24.

26.4 INSTALL A SUN SHIELD ABOVE THE CONDENSER

The cooling available can be increased by operating the condenser at a lower temperature. One way to reduce the heat of solar radiation is to install a cover above the condenser that will not interfere with its airflow but will prevent it from receiving thermal radiation directly. In effect this minimizes or eliminates the solar load.

26.5 SPRAY AN AIR-COOLED CONDENSER WITH WATER

Spraying water on a condenser improves the heat transfer rate on the outside surface of the coil, thus improving the overall heat transfer rate. The LMTD is also improved because the water temperature is usually colder than the air temperature. Further, the heat of compression is lowered due to the reduced condensing temperature. These factors improve the cooling capacity of the A/C system. At best this is a short-term solution, because water must be collected to avoid a mess on the ground nearby. On the other hand, this procedure fulfills a short-term need: when the area to be cooled becomes too warm for any reason, this is a way of immediately increasing cooling capacity.

26.6 PROVIDE COLDER WATER TO AIR-HANDLING UNITS

Suppose a chiller is designed to deliver 48°F water to air-handling units and the water is heated to 58°F in the exchanger(s). If the building's cooling load is increased 60%, one way of providing for the added cooling is to deliver colder water to the air-handling units. To illustrate, if the water delivered is at 42°F, the 58°F outlet temperature from the exchangers now represents a 16°F rise, equivalent to a 60% increase in capacity relative to the 10°F rise of the original design. An advantage of this solution is that the building's piping and pumps do not change, which helps reduce the cost while increasing the cooling available.

Chapter 27

Conditions That Can Make an A/C System Unacceptable

27.1 UNCLEANABLE A/C DRAINS

Moisture in air will condense on an evaporator surface over a wide range of operating conditions. Evaporators can be positioned horizontally or vertically or on any angle in between. Tubes must be arranged so that refrigerant migration and drainage can occur. A frequent problem is that drains are not located properly. They may be provided at different locations depending on the exchanger's orientation. A vertically mounted coil with tubes in the horizontal plane often has its drain located in the evaporator's lower side channel. This is ideal from a drainage point of view. The drawback is that if the drain should plug for any reason, there may be no reasonable way to clean it. In some buildings evaporators are located above the ceiling. The same problem can occur if the coil is installed horizontally. An evaporator drip pan beneath the coil has a drain at its center. When it is plugged water will collect in the pan and eventually spill, staining anything beneath it. In many applications no reasonable way for unplugging this drain is provided.

Recommendation: Engineers and checkers should confirm that the drain location is correct in horizontal and vertical units and that a way to clean or unplug drains exists.

27.2 NOISE

For a discussion of the subject of noise, see Sections 4.2 and 17.3.

27.3 RECIRCULATION

For a discussion of recirculation, see Chapter 20.

27.4 SCREEN ROOM COOLING CONDENSATION

A variation of the evaporator drain problem cited in Section 27.1 occurs in some screen and computer room applications. Cold air is supplied by the ECS and delivered beneath the raised floor of this room (refer to Figure 19-2).

The space between the base of the computer room or screen room and the building floor on which the screen room lies should be insulated. Otherwise, condensation will occur on this cold surface. As a result the ceiling will become water stained. If this region had been an actual duct, instead of equipment serving as a duct, the problem would be solved by insulating the duct. Painting ceilings and walls will temporarily correct the problem but will not prevent it from reoccurring. The problem is difficult to correct because it involves removing ceilings and working in limited crawl space. Any corrective insulation is likely to be installed several inches below where it would do the most good.

The ideal correction would be to remove the screen room and install insulation between the floor and the screen room base. The cost of doing this plus removing electronics and computers and reassembling and testing them, and the loss of operating time as the work is being done, make this option unrealistic.

27.5 SYSTEM SHUTDOWN DUE TO SOLAR RADIATION

Problems of solar radiation were discussed in Sections 18.10, 18.11, 21.10, 25.2, and 26.4.

27.6 EXCEEDING THE AIR VELOCITY LIMIT

The air velocity through a propeller fan (tip speed) should not exceed about 10,700 ft/min, to avoid excessive fan noise. The air velocity through a duct generally should not exceed 1200 ft/min to avoid duct noise (see Section 17.3 for a discussion of both issues).

PART 3

THE COOLING OF ELECTRONICS

Chapter 28

Electronic Installations and Their Heat Problems

28.1 NATURE OF THE PROBLEM

Communications, data, and information systems are assembled using various electronic components. Without the component, part or all of the system may be down, essentially a system failure. The most frequent cause is overheating (inadequate air flow area supplied) which historically occurs in about one of seven racks. Electronic cooling needs differ from those in the process and air conditioning industries. For the latter, the heat exchanger engineer selects equipment to provide the heating or cooling desired. They rarely select the components of the ECS duct system. Electronics engineers often approve the assembly of the screen room and duct system. Heat exchanger engineers become involved when overheating problems occur. In process cooling and in air conditioning , heat exchangers are usually the medium used to transfer heat. Generally (though not always), heat exchangers are not directly used in electronics cooling. The chapters of Part 3 describe typical problems faced in this industry and offer pointers on locating, avoiding, and correcting them. It should be noted that about 90% of heat transfer tasks in the process and air conditioning industries involve selecting exchangers and their components. In the cooling of electronics, almost all problems faced are related to solving and/or avoiding trouble jobs. No heat exchangers are usually involved except for the A/C system evaporator and condenser.

Electronics cooling presents problems not present in process or air conditioning exchangers. They are cooled by air using an ECS or A/C unit. The problem is that cold air is available at one location and cooling is needed at another. To move the air to where it is needed a duct system must be designed and assembled hopefully under a duct engineer's supervision. Duct system components, including the duct portion of screen room and racks, must be correctly sized so that sufficient air can be delivered to where it is needed. In systems that have an overheating problem, look for undersized flow areas in screen room and racks (a common cause of overheating). The larger flow areas needed in these engineered products should be identified before orders for them are placed. The cause of most electronics overheating failures is duct air flow areas, particularly in screen room and racks, that are too small.

As the first two parts of this book revealed, process and A/C heat exchanger problems usually have reasonable fixes except in cases where space is limited. This is also true of electronics cooling systems. The following are examples of some types of exchanger limiting conditions.

1. If a larger evaporator is to be placed in an existing A/C system, its size will be limited by the dimensions of the existing frame or by servicing needs.
2. The size of evaporator that can be built into an existing chiller is limited by chiller housing dimensions.
3. Generator air and hydrogen coolers are limited by the space available in the generator frame.

Similar limits apply in electronics installations. Reasonable fixes, once systems are built and assembled, are sometimes not possible, as evidenced by field observations of problem installations. These installations were cooling limited and would have made for simple solutions had problems been foreseen before systems were built. However, the problems became known after the systems were in operation. The following conditions were common to the problem installations.

1. The air-conditioning units were properly designed and packaged.
2. The systems were properly designed from an electrical point of view including component packaging, though duct sizes, rack mounting and spacing to provide cooling were not fully addressed.

A system that has been properly designed electronically may still fail if cooling needs are not met. Strange as it seems, in six of seven cases it is not difficult to correct problems in electronics cooling. Fans, properly positioned, can move cool air to provide cooling where it is needed. In the one of seven cases that present difficulty, the problem is usually a minimum air flow area that cannot handle sufficient cool air to cool the component(s). In most cases, the limiting flow areas occur in screen rooms and racks. In a sense, these are similar to the dimensions that limit the size of coolers in chillers, generators, and evaporator frames.

The engineer responsible for ducting should make a determined effort to identify potential problem areas before orders for screen room, ducts, and racks are placed. If done properly (nearly all problems are air-flow area related), almost all of the 200 or so overheating problems faced by the author over his career could have been avoided. Steps apt to produce favorable results are given in the paragraphs that follow. When facing conditions like those given in Tables 29-1 to 29-5, note that most problems could have been avoided had these conditions been followed before screen room, ducts, and racks were ordered.

1. The screen room floor height should be a minimum of 15 in high for a 10 ton ECS unit to avoid undersized ducting. The floor height should be higher for larger cooling units (Sections 28.7 and 29.2).
2. Only one and possibly two racks have several electronics boxes with high cooling loads (these are not easy to identify early in the project). These racks should be 4 in to 6 in wider than normal to obtain greater air flow area. The subject of "tight packaging" must be part of the assembly process (Sections 30.3 and 30.4). This is a best effort toward identifying the one rack in seven that will have an overheating problem.

Many smaller electronic boxes complicate this effort as they tend to reduce air-flow area and block flow.

3. In screen rooms divided into two rooms, be sure the exit flow area from each room and the return duct are 30 to 35% larger than normally needed. This allows available cooling to be transferred between rooms should it be necessary (Section 30.5).
4. If one has doubts about a specific rack's ability to deliver the cooling needed, place this rack at the end of a bank of racks, or leave space between racks, to install a wider rack should it be necessary.

Problems, not corrected by these steps, can usually be solved using one or more of the six methods given in Section 28.2. Section 28.5 also applies. The responsible duct engineer should confirm that the conditions noted are met before orders are placed for screen room, ducts, and racks.

It should be noted that electronics cooling problems almost always occur in equipment over which air-conditioning manufacturers have no control.

28.2 COOLING NEEDS OF ELECTRONIC INSTALLATIONS

Electronic units are often installed in racks to save space and for operating and maintenance convenience. Racks are used for mounting electronics in the computer, telephone, and aerospace industries, among others. Compact packaging of components, however, leads to a large concentration of energy in small volumes and is often the cause of overheating, hot spots, or failures. Cooling is needed to avoid such failure.

A/C systems or ECS units are often used to avoid overheating of electronic equipment in screen rooms, mobile vehicles, or racks. Selecting these products requires an engineer's services. In practically all cooling of electronic equipment, air is the heat transfer fluid.

Electronic equipment cooling loads are much smaller in scale than process and air-conditioning heat transfer loads. High operating ranges for the three categories are roughly as follows:

1. Power plant exchangers can transfer 500,000,000 Btu/h or more and have temperature limits to 2000°F and pressures to 15,000 psi (though they should not be subjected to both of these extremes simultaneously).
2. A/C systems may supply 2000 tons of cooling (24,000,000 Btu/h) or more at refrigerant temperatures below 400°F and pressures under 450 psi.
3. Electronic cooling loads typically range from 85 Btu/h (25 W) minimum to a maximum of about 12,000 Btu/h per rack at temperatures to 131°F and usually at ambient or airborne pressures.

The temperature change required of each varies similarly. Different solutions are needed to meet the needs of each industry. It does not follow that, because electronic cooling loads are smaller, they are less important. Providing for them is a must. Otherwise, equipment will fail, and systems as well.

Item specific heat exchangers are often used in electronic applications; only cool air is essential. Cabinets or racks designed to contain electronic equipment are built so the sides act as ducts to contain moving air. (Techniques for cooling electronic equipment in racks are illustrated in Figure 28-1). Often blowers 2, 3, 4, or 6 inches in diameter are positioned to move air to areas difficult to reach to provide the cooling needed.

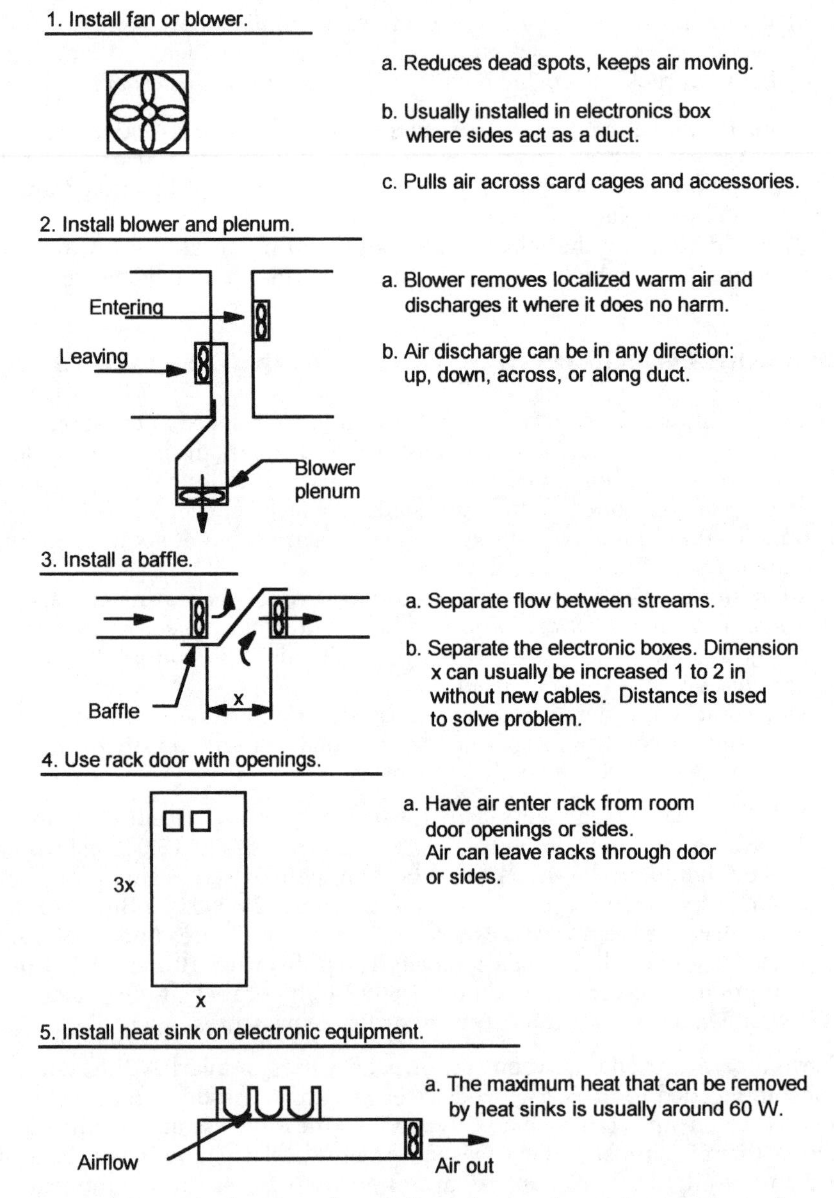

Figure 28-1. Techniques for cooling electronic equipment in racks.

Before an electronic unit is placed in a rack it is often bench-tested at room temperature for a week or more to determine whether it has an internal cooling problem of its own. If it performs over this period, it is considered to be adequately cooled in itself. Most components can sustain a temperature of 40°C (104°F) without failure. This temperature can be higher provided higher quality parts are used in their manufacture.

Bench testing at room temperature, however, is not necessarily a good predictor of behavior in racks. For example, cooling may be provided, as shown in Figure 19-2, by supplying cold air at the rack base; this air will leave at the top. In other designs the flow can be reversed with the cooler air entering at the top of a rack and exiting at its base. Cold air can also enter or leave at the front, side(s), or back of the rack. In closed and separate ECS systems, cold air usually enters a rack from one side, circulates past rack components, and exits from the same side it entered the rack. The air is continuously recirculated. It is independent of the cool air moved by the main ECS system. Equipment that has passed a bench test at room temperature may be installed near the top of a rack, where the air temperature might be warmer than room temperature—or internally, where component temperatures cannot be kept uniform. Thus, it could be subject to a local hot spot failure. The engineer's task is to keep the system on-line, temper hot spots, and improve reliability.

Another problem is that components are boxed in various sizes and shapes. One may be larger than another. If a larger unit is positioned beneath a smaller one and the flow is upward, the air supply to the smaller unit may essentially be blocked (see example 2 of Figure 28-2). Other factors can affect performance. For example, a larger box may have a lower cooling load, or any box could have a load low enough to be effectively zero. Individual units may have built-in blower(s) or none, and blower inlets or outlets can be at the top, bottom, or sides of an individual cabinet. Inlets and outlets can be on the same side. It is not uncommon to have two, three, or more components mounted on the same shelf with air leaving one component and going directly to the inlet of another. Though such a configuration is not recommended, because of the hazards of recirculation, units arranged this way sometimes work; success depends on the magnitude of cooling required in each unit. It seems reasonable to rotate electronic units 90 or 180 degrees to eliminate this condition, and it is feasible to do so provided that units are in position before cables are built and assembled. It is seldom a good idea after cables are made, because cables are not provided with slack other than what is needed for sliding drawers in and out. Cables are costly to assemble. They are rarely of sufficient length to allow equipment to be rotated; instead they generally need to be replaced with longer cables.

The designs most often used to correct overheating, hot spots, and other potential rack failures are as follows:

1. Use small diameter fans to move air to where it is needed.
2. A blower and plenum can be used to move warm air from an area (Figure 28-1, example 2).
3. Use baffles or existing rack channels to modify air movement at various rack locations (Figure 28-1, example 3).

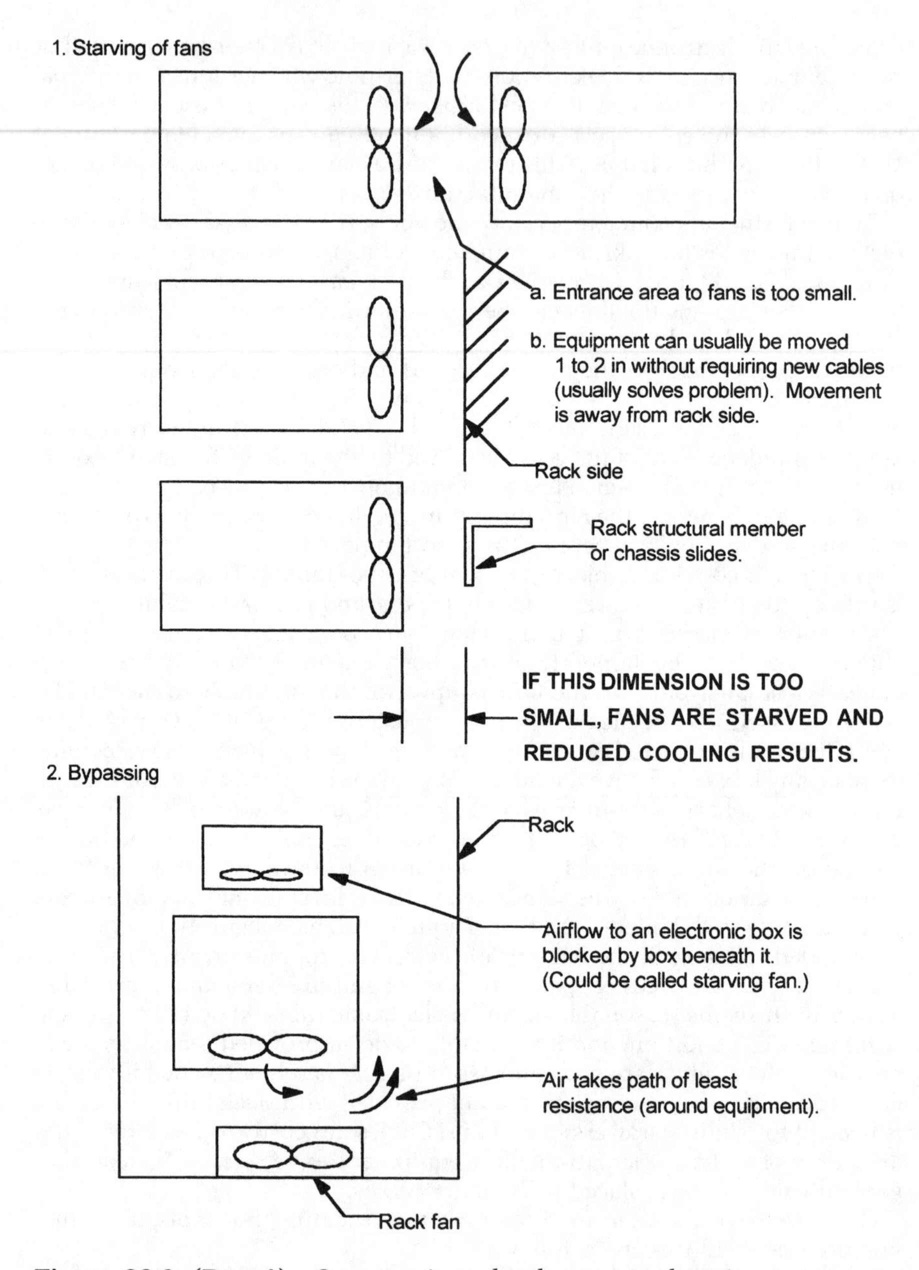

Figure 28-2. (Part 1) Constructions that have caused equipment to overheat in racks.

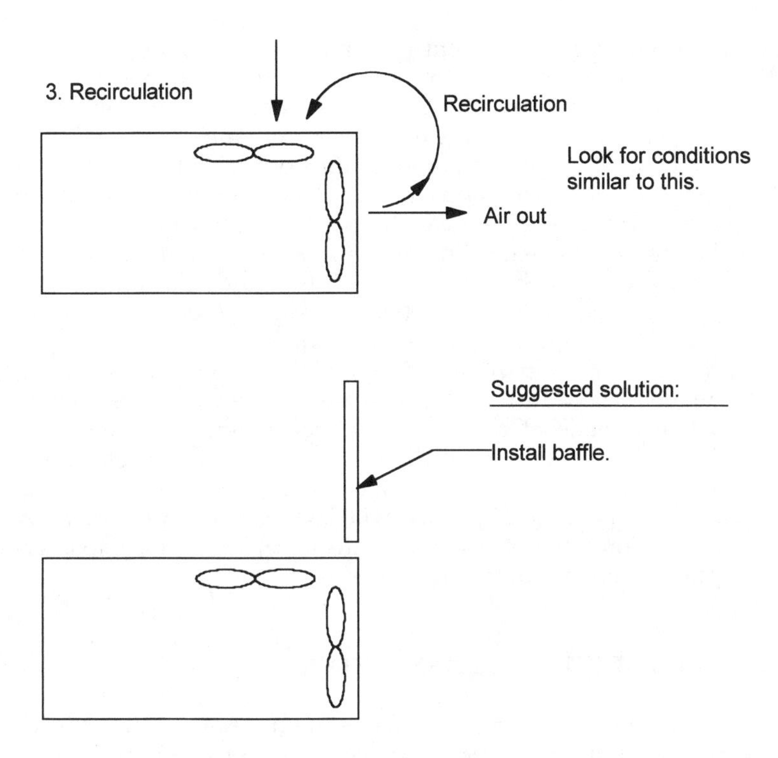

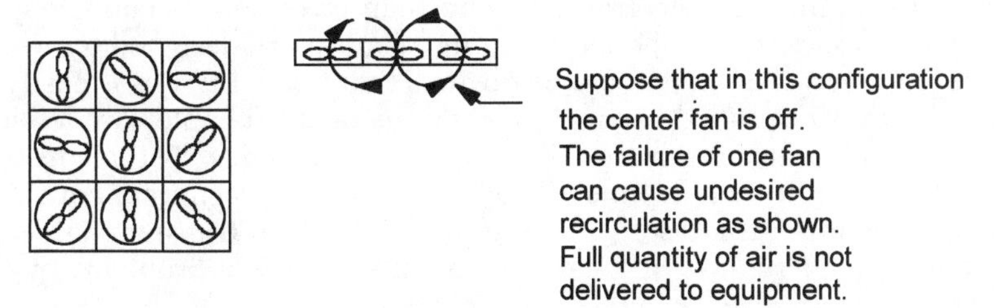

Figure 28-2. (Part 2) Constructions that have caused equipment to overheat in racks.

4. Consider using a packaged rack mounted A/C system where cooling is inadequate.
5. Provide a supplemental blower(s) near the top or bottom of a rack or both to increase flow through the rack (Figure 28-1, example 2).
6. Mount a heat sink(s) on the electronic box (Figure 28-1, example 5) which can temper hot spots.

These fixes are less costly to implement than are changes to installed electronics. They can be made in minimum time as well as minimum downtime. Using these fixes, electronics engineers avoid most potential overheating problems. When these do not work, heat exchange engineers are contacted. The first five suggestions are variations of the same idea, which is either to provide cold air (or more air) at electronic box inlets, or to design for removing warm air from near the box outlet(s) while preventing recirculation. The weak link is that just when more cooling air is needed, the remaining flow area is likely to be too small to allow sufficient cooling air to be provided. The common cause is tight packaging (Section 30.3) and large cooling loads resulting in too little area for airflow.

Heat sinks, the sixth of the suggestions, operate on two principles. First, they provide extended heat transfer surface to aid in removing heat. Second, their fins are usually made of aluminum or copper and thus they take advantage of the high conductivity of these metals. Assume an electronic housing has a high–thermal conductivity metal strip or heat sink attached. This metal, due to its good conductivity, will level temperature peaks by moving heat rapidly from the hot spot area, thus lowering the local temperature and reducing the chance of failure.

28.3 PLACING COOLING IN PERSPECTIVE

Racks and electronic units are highly engineered items. A great deal of work is needed to position equipment, design shelving, and locate chassis slides. Equipment cabling is made to specific lengths and requires a substantial amount of time to engineer and manufacture. It nearly always takes more time to engineer an electronic system than to design its cooling system. In the ideal case, the components and installation are designed to prevent failure from occurring. It should be obvious that fixing failures is costly.

The fixes described in previous sections of this chapter are less costly to implement than are changes to installed electronics. The changes can be made in a minimum of time as well as minimum downtime. For this reason electronic failures are sometimes addressed by heat transfer engineers. Electronic engineers, however, can also often solve these problems by trial and error using knowledge gained from experience.

28.4 TROUBLESHOOTING AIRFLOW PROBLEMS IN ELECTRONIC INSTALLATIONS

Two kinds of airflow problems can lead to overheating: recirculation and flow restriction. To troubleshoot an overheating problem, begin by confirming that the equipment performed during bench testing. Then ask, What is different about the current installation? Total the cooling load in the rack. Ask a second question: Is sufficient cool air being provided for the cooling needed? Observe rack components and their positions relative to each other. In most cases the equipment is installed near the front of the rack for easy access to control knobs. Cabling is packaged near the rear of the rack. The space between cabling and equipment may be an ideal bypass route. If excess cooling is not provided, a portion of the supply air can pass through a rack without providing substantive cooling. Sometimes, however, bypassing will not occur via this route: if a component's inlet blower is located near the rear of a rack, it is ideally located to pull in cool air.

Now consider airflow blockage. A particular problem frequently occurs when a vertical space is available in the front of a rack that is not occupied with electronic equipment (see Section 19.15). Here it is often convenient to install a rack drawer, and typically when a drawer is installed its base is perforated so as not to block air. But as soon as manuals or operating instructions are placed in the drawer, the documents block flow, a condition not wanted. However, such document storage will not necessarily present an overheating hazard where minimum cooling is required or where drawer lengths do not extend the depth of the rack.

The blockage of flow also occurs in many other forms (Chapter 19 detailed many that affect process and A/C systems). A common cause of problems is placing a blower inlet or outlet too close to nearby equipment, thereby reducing flow into or out of a unit. Another is to install two units so close to one another that insufficient air is delivered to either unit. Both are variations of starving a fan, except in this case it is the equipment or their blowers that are being starved. At several installations a condition has been observed in which space was provided for cool air to reach a unit but the exit from a nearby unit was located within that airspace, so that the unit was supplied with hot rather than cold air. A similar condition can be created inadvertently when manufacturers of electronic equipment, in a routine upgrade of their products, relocate blowers in an electronic box. The new locations may create a problem where none existed previously (see Chapter 19).

28.5 IMPROVING COOLING IN EXISTING RACKS

Consider a 19-in-wide, 36-in-deep, and 78-in-high rack. Flow will not be as smooth through this rack, acting as a duct, as it is through an actual 19- by 36-in duct. The rack includes structural members for supporting equipment, sliding shelving, bracing, cabling, openings with EMI/RFI filters, and other parts. These items and the equipment mounted on the rack shelving restrict airflow. Openings at the bottoms and tops of racks are usually 10 in by 10 in, 8 in by 8 in, or 6 in by 6 in or have flow areas close to these. It should be obvious that more air can pass through a 10 by 10 opening than an 8 by 8 opening and much more than through a 6 by 6 opening. Every component installed in a rack reduces flow area to some extent. When the flow area is too small to move sufficient air to components, flow can be increased by adding booster blowers at a rack's entrance or exit or both. In some but not all cases, these can deliver sufficient air to a rack where it is needed. When the flow area through the rack is small limiting air flow even though booster blowers are used, a serious problem exists.

Suppose a solid plate is positioned horizontally across a rack so that 90% of the flow area is blocked. The open area will be near the rear of the rack. Only a small portion of the flow will pass this plate, and it will be unevenly distributed in the direction of flow. Any component above the flow-blocking plate will not receive the cool air needed. Now, instead of the solid plate, imagine blocking part of this area with electronic equipment: the concept is similar. It is obvious that component positioning can reduce the quantity of cool air that can pass. The amount blocked depends on the remaining airflow area. Flow to equipment can be increased by installing blowers or by a combination of blower and plenum (see Figure 28-1). Baffles are often useful in directing flow.

If the flow area remaining after installation is too small due to tight packing (Section 30.3), a point will be reached where supplemental blowers will not deliver a sufficient quantity of air to cool equipment. The problem increases with increasing cooling load. The advantage of installing blowers, or blowers with plenums, or directional baffles in racks is that they are low-cost solutions to rack cooling problems, a major plus provided sufficient air can be moved. If they cannot move the air needed, the problem is a major one that may require redesign of the installation. The question might be asked, Why not build bigger blowers and plenums to move the air needed? This is seldom done, because the existing air flow area is usually too small.

There are many A/C cooling systems for rack mounting on the market. Sometimes adding one to a rack will solve a problem. A/C units can be installed on the sides or top of a rack. There are at least three major advantages to rack-mounted systems. First, the A/C unit can be positioned on the rack to deliver cool air exactly where it does the most good. Keep in mind that flow area from the side of a rack into the equipment is different from the vertical airflow area at the same position. The possibility of achieving adequate cooling through exact positioning depends on the magnitude of these flow areas and the quantity of cooling required. A second advantage is that these A/C units can be designed to deliver colder air than from the screen room ECS unit. A third is that they can, if necessary, cool one rack only, that is, operate independent of the central cooling system.

To illustrate why redesign of a rack installation can be a major problem, suppose components are tightly packed in a rack which is positioned at the center of a bank of seven racks. One solution is to remove the components and place them in a wider rack that has more airflow area. The added work entails any or all of the following:

1. Supplying new cables, each a few inches longer.
2. Purchasing a new rack and fittings.
3. Moving the adjoining racks a few inches. The relocation must not interfere with their air supply from under the floor.

4. Changes to other racks, particularly cabling, due to their being repositioned.
5. Determining that the new airflow area will be large enough to solve the problem of overheating.
6. Retesting of electronic circuits.
7. Taking the system off-line while these changes are made. Is the projected downtime acceptable?
8. Making sure labor is available to make the changes.

Such a scenario is not uncommon in electronics cooling. Electronic engineers, in the main, do not like the undertaking of changing to a larger rack. The author has seen two novel approaches to solving this problem. In one the user scrapped an electronic box and replaced it with an upgraded model that was smaller in size and used less power. This made more flow area available and reduced the cooling requirement, and the replacement component was available immediately. In another an electronics box was removed from one rack and placed in the adjoining rack. This reduced the cooling load in the first rack and thus the required airflow. Both were acceptable solutions. Refer to Sections 29.6, 29.7, and Table 29-3. Three possible solutions are referenced here. They were not used on any field problem known to the author but are possible solutions to other problems.

As Chapter 29 will repeatedly demonstrate, the general issues of adequate airflow in racks apply to screen rooms as well.

28.6 THE ELECTRONICS HEAT EXCHANGER SYSTEM

In process and A/C systems, heat exchanger manufacturers guarantee their selections to transfer the quantity of heat specified. Their responsibility covers their product(s) and extends from the face of the inlet nozzle(s) or ducting . Electronics cooling systems consist of an ECS unit, screen room, screen room ducting and racks. The ECS supplier is responsible for their portion of this system only. It is in the other components of this system (screen room, ducts, racks) that trouble often surfaces.

In shell-and-tube exchangers the manufacturer's product ends at the flange facings (or couplings) of both streams. Mating connections are the responsibility of the user. Similar reasoning applies to double-pipe, air coolers, and plate exchangers. For A/C systems, manufacturers usually supply packaged units , with the user responsible for piping to the unit (if cooled by water or a water/glycol solution) and ducting (if the unit is air cooled).

Companies that assemble electronics installations, on the other hand, do not always have a heat transfer engineer available to advise them. Thus, understanding some of the heat transfer issues becomes the electronics engineer's responsibility. Problems of tight packaging and undersized ducting similar to those described surface all too often. In the main, when they occur they are difficult and costly to correct. Most problems occur in highly engineered items that serve as ducts, or work in conjunction with ducts, such as screen room and racks.

28.7 ADDRESSING CAUSES OF FAILURES

The following is representative of the way electronics cooling systems are selected. Needs are defined by electronics engineers. ECS manufacturers design the cooling unit. Three components are needed to complete the system, screen room, ducts, and racks. All are part of the duct system. The author has addressed about 200 field problems and has not met a person responsibile for the design and performance of the duct system.

Ducts, or their equivalent, are needed to direct cool air to racks. Duct calculations for screen room and racks are rarely made. The author has seen one set in his career. Yet when heat transfer engineers are called to solve overheating problems, the corrective action, over 90% of the time, is some modification to the duct system. Two problems must be faced, locating the cause or causes of problems and providing for their correction.

The problems of bypassing, recirculation, or small flow areas (obstruction to flow) occur frequently in A/C and process applications (Chapters 19 and 20). They occur in electronic systems and may be difficult to correct because cool air must pass through the limited duct area of screen room and racks. Duct engineers seldom approve the positioning of equipment in racks which is often the cause of overheating. Nearly all problems occur in the duct portion of screen rooms and racks where A/C and ECS manufacturers have no control. Their responsibility ends at the connections to their equipment. Simply put, the design of ducts and equipment serving as ducts is not their responsibility.

Most screen room duct problems occur in designs similar to that shown in Figure 19-2. Racks and the area under the screen room floor are parts of the duct system. When electronics boxes are spaced too close to other units, air flow area is blocked. If too much flow area is blocked, trouble results.

Screen room features (some ducting is part of the screen room) that cause electronics overheating include tight turning radius or undersized elbows, a low floor height, small outlets, improperly sized ducts (see 400 and 550 cfm article, Sections 19.3 and 19.4), and good distribution. It may take considerable time to locate the source of the problem. Almost all are duct related.

Duct laws are usually not adequately considered when installing equipment in racks or screen room. Most systems will perform because supplemental blowers can provide the cool air needed. Less air is needed when the cooling required is low. There are limits to the amount of air that fans can deliver particularly if the approach area is small or the quantity of air needed is large. When these conditions occur, trouble appears.

28.8 ENVIRONMENTAL CONTROL SYSTEM CONCERNS

Consider the sizing of the environmental control system (ECS). The screen room may be in a building where the temperature is near 70°F. In this case there will be little heat transferred between the inside of the screen room and the area where it is positioned. Make-up air is delivered through the building's A/C system where moisture is removed. Hence most cooling needed is for lighting and electronics.

To provide the cooling needed cold air is delivered by an ECS system. It is unfortunate that the duct path often has flow limitations that occur in engineered items such as a screen room or rack(s) serving as ducts. Neither is easily modified without excessive down time and cost. The cause of overheating is usually easily defined but making the needed correction may be difficult.

An upgraded air filter may be added to the ECS cooling system to improve the quality of room air. Usually, the system designer will not be aware of this change. It may be made months or years after the cooling system was placed in service. This change will increase the duct system pressure loss, reduce the quantity of cool air delivered, and can cause a significant reduction in the cooling supplied by the ECS (Section 19.3).

A/C and ECS suppliers guarantee a given cooling load at a specific cfm of air and maximum pressure loss. However, undersized ducts and fittings increase system pressure loss and result in a reduction in fan or blower output thus reducing cooling capacity. Here are some of many reasons why ducts may be undersized.

1. To turn in a low bid and thereby sell a unit, a manufacturer may offer a marginally undersized unit and ducting.
2. The following is representative of a most irritating problem and one that is hard to correct. A duct system has been properly designed but then the building must be modified and requires additional steel at one location. The duct must then be modified to go around the new beams. It is almost a given that available space to make the change will be minimal. The result is a higher fitting loss than used in the calculations. The rerouting of the duct in this instance is similar to the modifications that had to be made to fit screen room ducting between the top of the screen room and ceiling above, in the example of Section 19.3.
3. A duct may have to go over or under another duct to avoid interference, and this adds to duct pressure loss.

Electronic overheating problems should be placed in perspective. Many heat exchangers (particularly repetitive ones) can be sized in less than an hour. Even two-fluid partial condensers can be sized manually in less than eight hours and in shorter times when computer programs are used. In cooling electronics the data needed includes that of many components, verifying duct dimensions, checking the needs in individual racks, flows, orientation concerns, recirculation in units, recirculation between units, and starving fans. It takes time to acquire this information, often a month or more to identify the cause or causes of the problem. It may take an equal time to correct.

Most electronics cooling problems are trouble situations that must be corrected. By comparison trouble occurs in only about about one in every hundred process and A/C applications. These estimates do not include electronics components that are designed to meet their own cooling needs.

Chapter 29

Frequent Causes of Electronics Overheating

29.1 OVERHEATING IN RACKS

Electronics overheating occurs in about one of seven racks. It can be caused by a small floor opening under the rack, a small rack opening, recirculation in and out of a component, recirculation between components, equipment spacing in racks, equipment blockage in racks, minimum air flow area, the size of booster fans, or an undersized rack. The best correction may not be available due to the possible lack of space for a wider rack(s). Providing for rack space, early in the design process, will simplify the solving of electronics overheating problems likely to be faced.

Duct details are rarely known to ECS suppliers. Many limiting conditions are built into the screen room and racks. Specific screen room and rack(s) requirements should be identified in the specifications. The challenge is to identify problem areas before racks and screen room are ordered and before the electronics system is finalized.

These circumstances also apply. The cooling system, ducting, air filter, rack sizes and locations are usually finalized long before the engineering of the electronic system is complete and tested. Cooling must be available as parts of the electronic system are completed. Substantial electronics changes may be made while the cooling system is being constructed. It is neither easy nor inexpensive to alter the cooling system after its assembly.

While late changes can cause overheating problems the important step is to build into the cooling system modifications that will minimize the "overheating of electronics". It is unlikely that every one will be identified as late changes can add to cooling needs while the capacity of the cooling system is fixed.

Major problems occur in about 15% of racks that serve as ducts. Consider the cooling of electronic boxes installed in them. Sufficient air must pass through the supply duct(s) or their equivalent. There is a minimum area that must be furnished; a lesser area is apt to limit flow causing overheating. Note the term, "minimum flow area" which is not the same as duct area.

Duct area is easily measured as is the flow area of an empty rack. Racks have built-in obstructions to flow that include mounting fixtures, sliding frames for equipment maintenance, cables, electronic boxes, inlet and outlet openings, and storage drawers to name a few. Simply put, these can reduce flow area and block passages so that sufficient air cannot reach the equipment to be cooled.

These problems can usually be met by using blowers strategically placed to move air to where it is needed. This is not a universal solution because racks with large cooling loads require more air than those having lower cooling needs. It is here that many overheating problems occur. Rack flow area is reduced if a large number of components are present as these tend to block more flow area. Remember that electronic engineers select the components that define the cooling load and determine where components are mounted.

To have a better understanding of overheating problems, the author's field experiences are cited. About one in seven racks has too small an air flow area to allow sufficient cool air to pass. About one in five ECS or A/C units are undersized for reasons given later in this chapter. Almost all screen rooms divided into two rooms have inadequately sized ducts for delivering the right amount of cool air to one of the rooms. To illustrate, assuming a ten ton system, each room is typically designed for five tons of cooling. In practice, the need is more likely to be 3 tons in one and 7 in the other or some other ratio. Flexibility to meet this need is seldom built in. One room's temperature may be at 64°F while the other is at 89°F due to poor distribution. At this condition the electronics temperature will be at a higher meaning equipment is more apt to fail. People entering the warm room know immediately that its cooling is inadequate. Undersized ducts add to pressure loss. These details are not known when screen rooms and racks are ordered. The result will be less air circulating. Of the dozen or so trouble jobs where flow measurements were taken, the air delivered in every case was less than that specified. These measurements were taken for designs similar to that shown in Figure 19-2. It does not follow that this applies to other screen room designs for reasons that follow.

29.2 OVERHEATING DUE TO SCREEN ROOM CONDITIONS

This observation applies only to the screen room design shown in Figure 19-2. A condition that can limit the quantity of cooling provided is a low floor height which results in a design that is similar to an undersized duct. It also is the cause of poor distribution when air is supplied to two rooms. A side effect is that this results in undersized elbows to the space beneath the screen room floor. This causes reduced ECS cooling and adds to the system pressure loss while reducing the quantity of air delivered by the blower (Figure 1-1). The author has not experienced reduced cooling from that shown on the nameplate in any of the screen room designs shown in Figures 19-3, 19-4, and 19-5 except when the ECS size was not well chosen. On the other hand nearly every screen room design similar to Figure 19-2 had an undersized duct system. Take note that rack related problems occur in all screen room designs, no exceptions.

29.3 THE SOLUTION TO A COMMON PROBLEM

Most overheating failures are caused by flow areas that are too small in ducts or in screen room and rack(s) that serve as ducts resulting in minimum air flow. Many conditions can cause failures. It is a time-consuming effort to identify which of the more than 50 reasons given in the Tables that follow is the cause of a specific failure. Despite this gloomy outlook, most overheating problems can be avoided by providing larger duct passages in the screen room flow path and in the one rack in seven that has this problem. An effort should be made to identify these problem areas before orders for screen room and racks are placed. Time spent on this effort will avoid nearly all the electronics overheating problems that are apt to occur if this effort is overlooked.

To assist in locating duct and other problem areas (including screen room and racks that serve as ducts) and to correct the problems identified, five tables have been prepared. These identify many problems experienced. The tables are a good reference to identify requirements that should be in the specifications. Problems specific to racks and screen rooms are listed separately as these items are furnished by different suppliers. The needed requirements should be in the specifications before orders are placed if overheating is to be avoided. The Tables are in these categories.

29.4 SCREEN ROOM AS DUCT CONDITIONS THAT HAVE CAUSED COMPONENT OVERHEATING (TABLE 29-1)

The principal causes of electronic overheating due to screen room conditions are given in this table. In Table 29-2 the causes applying to racks are given. In these two tables are listed the causes of most overheating problems. The point is to eliminate the conditions that cause overheating. In the flow pattern shown in Figure 19-2, reduced flow areas can cause problems for any of the first eight reasons listed in Table 29-1. Now refer to Figures 19-3, 19-4, and 19-5. Note that most of the reasons given are not a factor in these designs. None has a short turning radius elbow with a high pressure loss. When the return is under the floor, air can reach the return point from four directions, a low pressure drop condition. When air exits an ECS at the top, long radius elbows can be used. Small elbow exit areas are not a factor in the latter nor is distribution in the screen room. It is rare that duct flow area in these designs will be the cause of overheating electronics unless they were sized incorrectly. The floor height could be less using these designs but going this route may cause trouble later.

29.5 RACKS AS DUCTS CONDITIONS THAT HAVE CAUSED COMPONENT OVERHEATING (TABLE 29-2)

More electronics overheating problems occur in racks than at any other location. The purpose of the table is to identify and eliminate conditions that

Table 29-1 Screen Room as Duct: Conditions That Have Caused Component Overheating

1. **Elbow exit opening too small**
   The elbow outlet to the space beneath the screen room floor is often too small (Section 19.3). Low floor heights are apt to reduce the quantity of cooled air available. The flow area is often limited by screen room construction requirements. The only reasonable solution is to raise the floor height. The author is familiar with installations where the reduced size of this opening has reduced the cooling available by more than 25%.
2. **Flow impeded at outlet**
   This problem is easily identified by referring to Figure 19-2. The structural frames at the room outlet can reduce the flow of cool air.
3. **Distribution**
   In general, a unit—say, 10 tons in size—can be selected to deliver 5 tons of cooling to each of two adjoining rooms and perform as desired. It is not realistic to assume that at a future date, this design can provide 3 or 4 tons of cooling in one room and 6 or 7 in the other without building this capability into the system. When the screen room is no longer needed for its original purpose, it can be placed in a different service. A common occurrence is to put more equipment requiring cooling in one room than in the other. The result is often insufficient cooling in one room and excess cooling in the other. The A/C unit on hand is usually adequate in terms of cooling available, but the ducting, meaning screen room and rack, must be modified (a very difficult task) to take advantage of it. Another consideration is that much work has been performed on racks and equipment that are currently in place. The new requirements, when added to existing requirements, often exceed the cooling available to the room though not the system. It is best to have built-in screen room cooling transfer capability between rooms.
4. **Floor openings too small**
   The author has not met anyone claiming responsibility for sizing floor tile openings. Yet, some need to be larger. Once this problem is identified it can be an easy fix provided one gets permission to move the rack to have access to the floor tile. It usually requires this rack to be shut down while the correction is made. The best fix is to cut the desired opening in another floor tile and exchange tiles to minimize downtime.
5. **Ceiling opening and return duct too small**
   The connection at the juncture of ceiling and duct usually has a higher pressure loss than expected. Ceiling openings are usually square shaped and have the width of one tile. They are designed to be pleasant to look at and are normally furnished with a grille that has small openings evenly spaced. The pressure loss across the grille is far from negligible. This design along with an undersized duct can reduce the effective output of an A/C system by 25% or more.
6. **Duct return with short bending radius elbows**
   There is a limited amount of pressure drop available to move air through a duct system (Section 19.3). Short bending radius elbows almost always use more than is available. If, for space reasons, short radius elbows must be used, a larger duct return line and fittings will solve the problem. This is not as simple a fix as it sounds because it may require a signal leak test of the screen room and duct, which is costly.

(continues)

Table 29-1 *(Continued)*

7. **Expansion in the fitting too severe**
   To connect a relatively small return duct to the inlet of the A/C system requires an expansion connection. A large pressure drop will occur if expansion takes place in too short a distance. Space for this fitting is often the governing condition.

8. **Sound-deadening material added to duct**
   Sound-deadening material has thickness. When it is added to the inside of ducts, the flow area is reduced and the pressure loss in the line increased. This is another way that the available cooling is reduced.

9. **Floor and rack openings not lined up**
   It is not uncommon for the following scenario to occur. All racks in a bank have been ordered, received, and placed in position. Floor openings are made in the tiles so air can be supplied to the racks. Electronics engineers have more data available at this stage than when the rack order was placed and, with the new information, decide that one rack should have been wider. They replace this rack with a wider one. Other racks in the bank of racks are relocated a few inches and their openings are no longer in line with floor openings. This raises the pressure loss in the airstream.

10. **Undersized screen room ducting**
    This subject is discussed in Section 19.3 and in the preceding points 1 to 9. The author estimates that in 30 to 50% of all A/C and ECS systems where electronics underperformance was present, one or more ducts in the system were undersized. If an unexpected problem surfaces, such as a duct having to go around a column or another piece of equipment, one can be almost 100% sure this modification will result in an undersized duct and increase the pressure loss in the system.

11. **Filter changed to a HEPA type**
    The total pressure drop available through a screen room duct system is usually 0.5 to 0.7 in. HEPA filters, when fouled, have a pressure loss of near 1 in. The only way that these differences can be balanced is to reduce the amount of air flowing, which reduces available cooling.

12. **Electronics in mobile vehicles**
    An application similar to screen rooms is that of electronics installed in the back of mobile trucks. Because of a lack of space and headroom needed for ducting, it is not uncommon for ducts to be undersized, causing electronics positioned there to overheat.

cause rack electronics to overheat. Racks, installed in screen rooms, are usually ordered before the electronic system is finalized. The present concern is to provide sufficient flow area in racks so that cool air can reach components. Rack problems are independent of the screen room design chosen.

It is usually known whether or not an electronic unit performed during bench testing. If the equipment that has passed a bench test later fails in a rack, the cause is likely to be one listed in this table. Consider the causes listed as a group. A/C manufacturers and their heat transfer engineers have

Table 29-2 Racks as Ducts: Conditions That Have Caused Component Overheating

1. **Rack openings too small**
   Inadequate airflow area is provided to blowers that deliver air to and from racks. Be sure to account for obstructions at inlets and outlets.
2. **Direct recirculation**
   Air leaves an electronic box, returns to the inlet, and is recirculated (Figure 28-2 example 3).
3. **Recirculation between components**
   Air leaves an electronic box heated and goes to the inlet of another electronic box (Figure 28-1 example 3). The problem can be solved by using a baffle.
4. **Starving of fans and blowers**
   Inadequate airflow area was provided to several electronic boxes, preventing air from reaching some of them. One reason is that components are spaced too close to each other, blocking needed flow area. Another is that the flow area leaving the unit is too small, hampering exit flow (see Figure 28-2 example 1).
5. **Inadequate airflow area to a group of fans or blowers**
   The flow area to a fan may not include the added area needed when more than one fan or blower pulls air from the same inlet area. This problem also occurs when the air from several fans or blowers exits into the same area (see Figure 28-2 example 1).
6. **Dimensions of racks substituting as ducts**
   A convincing argument can be made that racks that contain overheating electronic components are too small because they have a limited airflow area. Minimum flow area to and from fans and blowers is the cause of most overheating problems. The same equipment placed in a wider rack with wider spacing and properly positioned blowers will solve most overheating problems. This conclusion is meaningless if the added space is used to place additional equipment on the shelves of wider racks without a corresponding increase in spacing between components (see Sections 30.3 and 30.4).
7. **Bypassing**
   Blowers must be positioned to minimize bypassing in the rack.
8. **Airflow blockage between components**
   Large electronic boxes can block flow to smaller ones (see Section 28.2 and Figure 28-2 example 2).
9. **Flow-directing devices**
   Earlier fixes (usually the addition of baffles) were implemented and later removed by others who did not understand their purpose (Figure 28-1 example 3 and Figure 28-2 example 3).
10. **Increased cooling load**
    An A/C unit that provides cold air may be undersized due to equipment upgrades that increased the cooling load. Upgraded equipment in racks also changes local cooling requirements. Most problems identified in this table could potentially be altered by this change.

(continues)

Table 29-2 ***(Continued)***

11. Local airflow in rack unfavorable
Airflow can be reversed locally, usually by installing a blower to move air in a direction opposite that flowing in a rack. There is nothing wrong with this solution. It works most of the time (Figure 29-2, example 2).

12. Nonelectronic devices affecting flow
Installing a rack drawer will modify the flow and block airflow to some degree in the rack (Section 19.15).

13. Undersized screen room ducting
The ducting that delivers cool air to racks is undersized or has components with high pressure losses. This reduces the quantity of air that can be delivered to racks (Section 19.3 and Table 29-1). It lowers the cooling that can be delivered by an A/C or ECS system. Electronic engineers often claim cooling provided to electronics is too small. In a sense they are right. The reason for the shortfall is nearly always that the duct system that carries the cool air is too small even though the cooling unit is properly sized. The conclusion is correct but for the wrong reason.

A rule of thumb in A/C work is that 400 cfm of air equals one ton of cooling. For ECS units, this number is nearer 550 cfm because the heat of condensation was removed by the building's A/C system. It is the author's opinion that undersized ducts are often the result of engineers' mixing these values. The amount of undersizing is frequently close to the ratio of these values.

14. Location within rack
At some rack positions, the local temperature may be higher than room temperature. Components are usually tested at room temperature. This is sometimes the cause of overheating.

15. Tight packaging
This is probably the number one cause of electronics failures (an undersized rack often causes the problem). It is usually difficult to correct due to lack of airflow area (Section 30.3).

16. Rack fixtures
Installed rack fixtures (frame, shelving, sliders, etc.) change airflow patterns.

no control over the rack's selection, the packaging of electronic equipment or a component's position in a rack, bypass areas around equipment, duct and fitting sizes in A/C systems, or filter changes. They do not approve facility design features that obstruct flow in racks and ducting. The only possible cause of overheating controlled by the ECS manufacturer is an undersized evaporator, and this is a rare occurrence. Yet users almost automatically assume the ECS unit is the cause of the failure. It rarely is. Tight packaging, which blocks sufficient cool air, is the most frequent cause of overheating failures.

29.6 SUGGESTIONS FOR CORRECTING EXISTING ELECTRONICS OVERHEATING PROBLEMS (TABLE 29-3)

This Table addresses problems that exist in undersized racks or in newer racks where flow area has not been adequately addressed. *It itemizes recommendations for solving existing problems.*

The following should probably be added to Table 29-3. The suggestions that follow were not included because they were not field problems or solutions that occurred but these can be solutions when flow is limited. Assume a rack has several shelves. The shelf with the largest obstruction is the limiting flow condition. Why not shift components between shelves so each shelf has a total obstruction of about the same value. In this way the flow area will be increased allowing more air to pass. An acceptable variation would be to relocate an existing electronics box without the need to change its cabling. The new position will be limited by the existing cable length and the box may be above or partly above another box. The idea is to make available more air flow area. Supporting the box in the new position should not be difficult. A potential drawback is sliding the rack drawer in and out without interference between electronic box and rack frame, a condition that may have an easy remedy.

Another possible solution follows. Assume two (it could be any number) electronics boxes have internal blowers and air enters and leaves in a horizontal direction. Usually these boxes are mounted on a shelf and each blocks air flow area. Why not mount them on top of one another. In this arrangement, there is less air flow blockage and more area for air flow, exactly what is wanted. Before proceeding make sure that the hot air out of one box does not go directly into the inlet of another which could occur at either box.

A third option is to replace a given rack with one of the same width except that the rack will be taller than any other in a bank of racks. If a component is removed from one of its shelves (the component removed will open flow area on this shelf) and placed on a new higher shelf in the rack, the end result will be a larger air flow area.

29.7 SUGGESTED CHANGES IN SCREEN ROOM AND RACKS TO AVOID ELECTRONICS OVERHEATING (TABLE 29-4)

In this table are listed reasons why electronics overheat. Corrections to solve these immediately follow. The correction in most cases is to enlarge the duct or flow area locally. Not all racks are effected, only about one in seven needs to be wider. The table presents the best argument for raising the screen room floor height to 15 in. in systems that have external ducting and 10 ton ECS units. Higher floor heights are needed for larger units. Note that this recommendation applies only to the screen room design shown in Figure 19-2 while rack conditions apply to all racks regardless of the screen room design chosen.

Table 29-3 Suggestions for Correcting Electronics Overheating Problems

In existing racks

1. Substitute another electronics component that performs the same function as one you have but is smaller in size and hopefully uses less power. Ideally, this should increase the airflow area.
2. To reduce the rack cooling needed and make more flow area available, consider placing one or more components in an adjoining rack where more cooling is available.
3. Disassemble an electronic box and repackage it so components block less airflow area.
4. If space is available, use a larger blower to direct more air to where it is needed.
5. In some cases overheating can be corrected by using heat sinks (Figure 28-1 example 5).
6. If space is available add a side-mounted A/C system to the rack. Remember that cool air traveling horizontally through a rack has a different airflow area from air moving vertically.
7. Rearrange the components so that the one requiring the most cooling is positioned where the rack air is coldest.
8. Check that no component receives its own recirculated air (Figure 28-2 example 3).
9. Check that air leaving one component is not going directly into another (Figure 28-1 example 3).
10. Check that structural members or other components are not positioned to starve fans or block an air exit (Figure 28-2 example 1).
11. Establish from previous jobs the minimum distance that should be maintained between components. Above all, maximize airflow area. Tight packaging is almost a guarantee that electronics overheating will occur.
12. Repeat the process if a rack component is replaced with another.
13. Not all electronics boxes must be mounted horizontally. Some can be mounted on a side, blocking less airflow.
14. Install a baffle to block or direct flow (Figure 28-1 example 3 and Figure 28-2 example 3).
15. Leave space for a wider rack. The finished system may require a larger airflow area.
16. Confirm that inlet and outlet areas to the rack are adequate.
17. Move air to where it is needed using any or all of the suggestions given in Section 28.7.

In new racks (see also Table 29-2)

1. Decide if one or more racks should be wider (Sections 30.3 and 30.4).
2. If possible, provide a 4-in clearance around all electronic boxes, particularly where recirculation is apt to occur.
3. Provide some space in a bank of racks so a wider rack could be installed, if needed. A rack that should have been wider may have been overlooked (Sections 30.4 and 30.5).
4. Use applicable suggestions from the 17 listed in the first part of this table.
5. In racks with high cooling loads or many components, point 1 in the second part of Table 29-3 is almost mandatory.

Note: See Section 29.6 for other possible solutions to overheating problems. None of the three examples given were field problems but all are potential solutions for them.

Table 29-4 Suggested Changes in Screen Room and Racks to Avoid Electronics Overheating

Most duct system corrections are screen room and rack corrections which should be made before a system is ordered, not after. Almost all difficulties are caused by insufficient flow area for air.

Screen room as ducts: corrections

1. ECS exit elbow should have same inlet and outlet flow area.
   Correction: A floor height of 15 in for a 10-ton unit will usually satisfy this condition. A higher floor height is needed for larger ECS units.
2. Sufficient flow area is needed to allow good air distribution to two rooms.
   Correction: A floor height of not less than 15 in for a 10-ton unit (increase the floor height for larger units) and larger screen room outlets will usually satisfy this condition.
3. Screen room outlets have high-pressure drop grilles for aesthetic purposes.
   Correction: Enlarge screen room outlets or add a second outlet from each room.
4. Floor openings to rack(s) are too small.
   Correction: Furnish the floor tile with a larger opening. Align tile with rack opening.
5. Quantity of air furnished is less than specified.
   Correction: Have screen room supplier confirm that data sheet values are being delivered. Confirm that ducts were sized for approximately 550 cfm of air and not 400 cfm as air moisture is usually removed in another A/C system (Section 19.4).
6. Reduced airflow.
   Correction: Confirm ducts are not undersized and the filter does not affect blower output.
7. The wider rack need cannot be installed.
   Correction: Leave space for a wider rack should it be needed.

Racks as ducts: corrections

1. A wider rack is needed to avoid electronics overheating.
   Correction: Identify the rack that should be wider (difficult to identify, early decision). May require changes in rack spacing, cable lengths, relocation of other racks, repositioning floor openings, a new rack, disassembly of equipment in the old rack and reassembly in the new one.
2. Temperature in room where racks are installed is too high causing rack electronics to fail.
   Correction: See point 2 above. Also check if ECS or A/C unit and their ducting is undersized.

29.8 CAUSES OF ELECTRONICS OVERHEATING FAILURES OTHER THAN DUCT CONDITIONS (TABLE 29-5)

This table is a list of conditions, other than duct size, that can cause electronics to overheat. They should be factored into the design before the ECS system is ordered. Future problems that might be avoided are high relative humidity and lack of access to repair a unit. The latter often occurs in mobile units but rarely in fixed installations. When needed requirements are in the specifications, most problems can be avoided.

Table 29-5 Causes of Electronic Overheating Failures Other Than Duct Conditions

The author's experiences in cooling of electronics are mainly in one-of-a-kind applications. The causes of failures apply only to the screen room and its cooling system and ducting, or to trucks and trailers where cargo space was used for similar purposes. All problems occurred. These are possible causes of electronics overheating.

Undersized A/C or ECS units: A/C and ECS units are similar to heat exchangers in that the cause of inadequate cooling is seldom a poor engineering selection. This said, cooling units are sometimes inadequate for reasons that follow. Keep in mind that it is relatively easy to upgrade electronics systems but it is not easy to upgrade most electronics cooling systems.

Added cooling requirement: An existing electronics system is upgraded by adding three new components. Two problems must be addressed. If the components are placed in the same or in other racks, will the cooling in these racks be adequate? An equally important question is, Will the A/C or ECS system in place be large enough to handle the increased cooling required?

Undersized for new applications: Sometimes screen rooms are not needed for their original purpose. They are available and could be used on other projects. Under these conditions the built-in cooling system may not be adequate though the screen room space may be ideal for the new conditions. When seeking the cause of inadequate cooling, ask if the screen room was chosen for its current use.

Air filter upgraded: The design pressure drop in the A/C or ECS system, including the air filter, may be 1 in. A change to a more efficient air filter may increase the system pressure loss to 1.3 in. or more. Because of the higher pressure loss the fan or blower will deliver less air (in effect less cooling). Thus, the cooling system is effectively undersized because the duct system is undersized.

Undersized ducting: An undersized duct system will increase the system pressure drop just as an upgraded air filter does thus decreasing the available cooling. Because less air is circulated it sometimes causes an undersized cooling system. This problem often occurs in mobile systems.

Inadequate cooling in one of two rooms (distribution/divided flow): When the cooling available is inadequate, equipment may become warmer but not necessarily fail. Room temperatures may rise to 78°F or more making personnel uncomfortable, which is a type of failure. The A/C engineer will hear of it. The usual cause is poor distribution (not enough cool air is delivered to the room) or an undersized A/C or ECS system (Section 29.1).

Oversized A/C and ECS units: An oversized A/C or ECS system introduces problems of a different sort. The next two items are two of them.

High or low room RH: If a room or several rooms have too large an A/C or ECS unit to provide cooling, the unit may be on less than 5 minutes per hour. This is not enough time for all room air to pass through the cooling coil. Under these conditions little moisture is removed at the evaporator, resulting in a high room relative humidity (Sections 22.2 and 25.1.2). If the RH must be lowered, consider adding heaters to activate the ECS so that it will be on longer to remove moisture at the evaporator.

(continues)

Table 29-5 ***(Continued)***

Room too cold: Rooms cooled using an oversized A/C or ECS system will be too cold in winter unless an alternate source of heat is available. Typically, these rooms have high relative humidities because they are not on most of the time.
RH too low in winter (less than 50%): This problem is easily solved by adding a humidifier of the proper size which will increase the room RH. If the RH must be below 50%, installing a desiccant dehumidifier should solve the problem.
Access to A/C and ECS units for service or repair: A/C and ECS units associated with screen rooms nearly always are provided with sufficient space for servicing or repair. This is frequently not true of units designed to cool electronics in mobile vehicles. Two common errors follow.
Location of controls: The operator should have easy access to controls in all mobile cooling systems. ECS and A/C refrigerant piping should be assembled so that controls are located to make it easy for the operator to make system adjustments. Mobile vehicles access to repair: The main limitation in mobile vehicle A/C and ECS systems is lack of space. Therefore, it is essential that procedures on how repairs will be handled should be finalized before units are ordered and installed.
Compressors: The cooling systems on mobile vehicles must be reliable. A not uncommon problem is that the compressor selected is not designed for over-the-road conditions, notably shock loads, which can cause compressors to fail. Make sure the compressor selected is designed for this purpose. Compressors of this type are readily available.
Compressor heaters: Compressors in mobile units experience wider operating temperature ranges than stationary units because they operate in many different climates. To start in cold weather, they must overcome the effects of refrigerant migration. They need a reliable built-in oil heater to avoid start-up problems. Heaters sometimes fail. It is a good idea to check for electrical continuity at regular intervals to make certain heaters will perform when needed. Compressors frequently fail due to refrigerant migration particularly if the heater is not working.
Shutting down over a weekend: The following subject is discussed in Section 23.3.3.
Undersized evaporator: Suppose a unit was designed for a 20°F temperature rise in the racks. The temperature of the air off the evaporator coil is 55°F and that leaving the racks is 75°F. Three conditions should be noted. The air off the racks (meaning most of the air circulating) enters the room at 75°F. The actual room temperature will be below this value due to incoming 55°F room air. A second consideration is that the room temperature should be 70°F. This means that an increase in the quantity of room air will be needed to temper the racks' exit temperature. This condition becomes worse with increasing temperature out of the racks. A nonstandard ECS will be needed to deliver it. It means that the duct size and evaporator face area must be increased. Assume a 75°F room temperature is acceptable. If the evaporator is undersized, even though the specification sheet says the air temperature leaving is 55°F, the actual temperature off the coil may be 62°F, in which case the temperature out of the racks would be 82°F, clearly unacceptable. Space may not be available to install the larger evaporator needed. The lesson is that specification sheets and performance curves supplied by manufacturers are not always correct. Undersized evaporators are difficult to detect unless one has performance curves of other evaporators known to be accurate that can be used for comparison purposes.

29.9 FAN AND BLOWER CONSIDERATIONS

Electronics are cooled by air to prevent them from being overheated. Cooling occurs at component surfaces; the heat transfer rate can sometimes be improved by the proper use of heat sinks. The heat load is the electric power measured in B/hr. Recall that in air coolers, large flow areas are needed to move air across a few tube layers. Similarly, relatively large flow areas are needed to direct atmospheric air to electronic boxes. If the components are packed too close to one another, the flow area to fans or blowers, including those built into equipment, will be minimal thus creating a starving the air mover situation preventing cool air from reaching the electronics.

Fans and blowers can be starved if *too small an inlet flow area is provided between electronic boxes*. Tight packaging creates another condition essentially unknown in the process and air conditioning fields. Starving fans can also occur *if the exit area between boxes is too small preventing air from leaving*. Either or both conditions, in the same installation, limits cooling.

A great deal of time and engineering goes into the design of electronic systems but usually not in keeping them cool. Cooling problems surface after units have been in service. An air distribution problem can exist before even one component is installed in either product, that is, the screen room floor height may be too low (minimum area for distributing air) or rack inlet and outlet areas may be too small. The problem is that electronics engineers too often mount components in racks with little consideration given to equipment cooling needed. Fans or blowers can help but not in all cases.

Rack air flow can be modified to some extent by positioning blowers, blowers with plenums, baffles, flow channels, mounting A/C systems on racks where needed, adding heat sinks, avoiding bypassing, adding extended surface, reversing air flow locally or taking steps to avoid starving fans and preventing recirculation. These corrective steps are not part of the ECS unit. They occur in areas and products where ECS manufacturers have no control.

Overheating becomes known when adding blowers does not solve the cooling problem. To make corrections usually means shutting down, repositioning electronic boxes (provided there is sufficient space which may not be the case), new cables, and testing both electronics and cooling. In some heat exchangers, seal strips are used to direct flow across surface. Racks usually have large bypass areas and seal strips are not practical. Instead, blowers are used to direct air locally across heat transfer surfaces to exchange heat. Blowers, properly positioned, will minimize overheating problems.

Chapter 30

Overheating Electronics and Corrective Steps to Take

In A/C and process applications, the problems of bypassing, recirculation, or small flow areas (obstruction to flow) occur all too frequently (Chapters 19 and 20). Similar problems occur in racks and screen rooms. The corrections needed may be difficult to attain because the cooling supplied by the A/C system must be delivered in engineered systems (screen room and racks serving as ducts, and blowers placed where needed) over which the A/C supplier has no control. Those that assemble equipment in racks seldom have a heat transfer engineer available. Even if available, electronics engineers would seldom allow a change in the electronic system when the odds are 85% to 15% that their design will work with a good blower(s) selection.

Racks with large cooling loads require more cooling air than those with lesser requirements. Electronics engineers select the equipment that defines the cooling load. An input often overlooked is that of the duct designer. Ducts, or their equivalent, are necessary to carry the cool air from the A/C unit to racks where the heat of electronics is removed. Duct calculations are rarely made. Yet, the corrective measure is almost always to modify the duct system in some fashion. Ignoring duct engineering is the cause of most electronics failures. It is not surprising that duct problems are not addressed by the A/C supplier as their responsibility ends at the connections to their equipment.

Part of the cooling system is the rack sides and the space under the screen room floor which serve as ducts. Electronic boxes installed too close to each other in racks block air flow. This is not always true but it is in many trouble situations. Most racks will perform because supplemental blowers will provide the air needed to cool equipment. It is when the limits of these fans are exceeded that trouble surfaces.

30.1 A SCREEN ROOM RECOMMENDATION

The author has not had the experience of selecting a screen room, its racks, or an ECS unit for cooling electronics. However, he has had to correct the overheating problems of over 200 electronic components that were installed mainly in racks. About 90% of the failures were the result of undersized duct system components. Here are possible causes of reduced cooling.

- The as built size of screen room duct components was smaller than that used in the calculations.
- The elbow area to the space under the floor is too small

- The floor height is too low (the air flow area under the floor is small)
- The room outlets are small
- Floor and rack openings are not lined up
- Electronics are packaged too tightly in racks (Section 30.3 and 30.4)
- Space is not available to make needed corrections to increase flow area
- More cool air is needed than a rack (acting as a duct) can handle
- An air filter was upgraded to a more efficient one reducing the quantity of cool air that a fan or blower can deliver.
- Air recirculates to a component or between components
- Starving the air mover due to small air flow areas at the inlet, outlet, or both.
- An inadequate ECS size (See Table 29-5).

Note that all reasons are directly or indirectly related to the duct system or to racks and screen room that act as ducts. It is the author's view, based on field experiences that most electronics overheating could be avoided if designers paid more attention to air flow areas in the screen room and racks. This includes elbows, tees, connection details, duct dimensions, rack widths, and the increased air flow areas needed for large cooling loads before they sign off on the equipment to order. Screen room cooling problems are usually not a room's size but the duct flow areas that supply them. Floor height is usually a factor. The duct choice may be poor one because cooling needs are not finalized until later but this is an unlikely error. Electronic systems are updated as construction proceeds which can effect the cooling required of the system.

There is little that can be done to correct an undersized cooling unit. Replacing the ECS or A/C unit with a larger one will not solve the problem as ducts sizes must also be enlarged. To minimize electronics overheating problems, the author suggests the following screen room changes be made in designs with external ducting and flow under the floor similar to that shown in Figure 19-2.

The minimum height of the screen room floor should not be less than 15 in for a 10 ton unit particularly when the screen room is divided into two rooms. The author is aware of screen rooms that perform at a lower floor height mainly because the cooling required is low. The installation that was undersized (65% of cooling delivered, Section 19.3) performed 85% of the time because the cooling needed was below the design cooling load most of the time. However, when fully operational it did not perform. Raising the floor height and increasing the size of outlet will avoid the problem (Table 29-4).

Screen room ceiling duct openings in two-room screen rooms should be at least 50% larger than those currently in use for several reasons. One is to allow about 40% of the cooling available to be transferred from one room to another. This is near the practical limit of cooling imbalance between rooms. Another is to reduce the resistance of dampers, best located in the return duct, if balancing flow between rooms is needed. The larger flow area will accommodate the grilles with small circular openings that are often used for aesthetic purposes. These have a high pressure loss and large openings provide some flexibility. If the larger exit area is not needed for distribution the reduced pressure loss allows for some upgrade of air filters. The reduced cooling capacity described in Section 19.3 will be largely avoided and the limited built-in flexibility may be valuable on future applications. The outlet duct should be the same size its entire length as this distance is short. In total these recommendations provide:

1. More cooling flexibility between rooms
2. The chances of available cooling being reduced are minimal
3. The chances of the screen room duct flow areas being too small (there is no other means of correcting this problem should it occur) is minimal using this procedure
4. It provides flexibility for a future use of the screen room
5. Historically, the author has not experienced an electronics overheating problem attributable to properly sized screen room ducting. This is not true of undersized ducting in other parts of the system. These comments do not apply to racks which are discussed separately

These changes are relatively minor and inexpensive when compared with the time and cost of correcting electronics failures. The purpose is to reduce failures. Probably the main reason for recommending them is that, if it is known that larger flow areas are needed after units are built, assembled and operated, there is no reasonable way of making corrections to the now existing installation. These recommendations should essentially eliminate the problem. Other pluses follow. A higher floor height should allow the inlet elbow to have the same flow area at its inlet and outlet. Cable blockage (Figure 19-1 example 2), for example, would likely be minimal if the air inlet opening is increased. A similar kind of obstruction occurs in mobile units when no elbow is needed. The ECS outlet is direct connected to the space beneath the floor. Structural frames impede flow the same way as cable blockage. If, using a 10 ton ECS unit, the cooling required in one room is 7 tons rather than 5, the increased floor height will allow air to flow freely under the floor to where it is needed. In installations where the cooled area is divided into two rooms, larger ceiling openings allow cooling to be shifted from one room to another. This may require that separate fans be provided to move air from beneath the floor to one room.

Once a screen room design is finalized, make sure there is insulation between it and the floor to avoid condensation on the ceiling below (Section 27.4).

30.2 AN OVERVIEW OF SIZING DUCTS AND RACKS

The cooling load in most racks is small enough and their air flow area large enough that electronics cooling is easily accomplished by the proper positioning of blowers and baffles. This section addresses the one in seven racks that has a high cooling load and less than adequate flow area. An electronic unit that fails can cause the system to fail. Hence, the solution is critical.

Electronics engineers control the design of screen rooms and racks and make most of the decisions noted in this article. The intent is to identify designs that result in electronics overheating. Construction details are usually the cause. The conditions are rarely known to the A/C equipment supplier. If electronics overheating is to be avoided, it is a must that the recommendations be followed.

Electronics cooling is strongly influenced by products that must be available before the cooling load is finalized, an unfavorable circumstance. Included are screen room, racks, A/C or ECS systems, ducts, size of inlets and outlets, and distribution to name a few. The screen room and racks must be on hand as electronics are assembled. These are part of the duct system. Ducting (particularly screen rooms and racks acting as ducts), though changes needed are simple before assembly, are not easily modified after a system is assembled. Changes in duct details usually effects electronics systems components that have been assembled. If correct decisions on screen room and rack dimensions are made before an order is placed and before electronic boxes are installed, overheating problems can be reduced to a minimum. The decisions include which rack(s) should be wider (usually those with high cooling loads and/or many electronic boxes), what their width should be, will floor space be available for a wider rack(s), will a review of jobs that overheated be made to avoid repeating earlier errors, will the flow area beneath screen rooms and racks be adequate, is tight packaging avoided, and will the air filter needed be of a higher quality than standard. This is a tall order as the cooling required, the desired rack width, and the preferred spacings must be made without complete data. The author's experience is that adequate air flow area in screen rooms and racks will solve nearly every overheating problem likely to be faced. In other words, the suggestions for the construction of new facilities are similar to those that will be needed to solve existing overheating problems except that, if the suggestions are implemented before an order is placed, component overheating can be avoided. The effort is to make these changes before ducts and electronics are assembled. To do this, it is essential that tight packaging be understood.

30.3 TIGHT PACKAGING

Assume a rack contains electronic boxes and has a sufficient flow area to allow air to remove a quantity of heat, "Q". Now assume another rack of the same dimensions and free flow area except that the heat to be removed is "6Q". In the first example the air temperature may rise 10°F in passing through the rack satisfying the cooling load, "Q". In the second rack the same quantity of air will have to cool "6Q". The temperature rise in this rack should be six times as great or 60°F. If the air entering the rack was 55°F, the leaving air will be 115°F. (55 + 60°F). To reach 115°F, the electronics will have to be above 115°F. The upper temperature limit of many electronic boxes and components is 40°C (104°F) so failure is almost assured.

The problem may be further complicated because the air entering temperature may not be 55°F but 75°F due to the position of the electronic box in the rack.

Consider the options for avoiding these failures. One is to use one or more blowers to double the amount of air passing this section in the rack. If the flow were doubled the temperature rise would be 30°F which would probably solve the problem. The same result is obtained by increasing the air flow area and spacing between components. The variables of blower selection, spacing between components, and heat load can be balanced to avoid overheating in six of seven racks. In the seventh case, the heat load is usually high, spacing between components is tight, and blowers cannot move the needed flow through the small areas available. Unfortunately, in some racks the flow area cannot be increased because all available space is in use including that for a wider rack. The lack of flow area may cause a component failure and a system failure as well. The airflow area needed depends on the amount of cooling required. It also depends on the resistance met in other shelves. The quantity of air can be increased locally by adding a blower or a combination of blower and plenum, but there are limits on how much this this can increase cooling. Do not consider a problem solved until the effects of rack internal recirculation, if any, are factored in. Higher cooling loads require larger flow areas. A possible correction is to reposition the equipment in the rack. However, there are several reasons why this may not be an acceptable solution.

1. There may not be space for it.
2. It takes time to implement this recommendation, meaning downtime.
3. The cost of making the change(s) may be excessive.
4. The rack may be in a bank of racks, thus compounding the problem.

This problem can be avoided in nearly all cases if the one rack in seven rack selected had been wider assuming all racks have the same depth. If one made this choice initially, installed a wide rack but ignored "tight packaging," using a wider rack is probably useless.

30.4 THE CHOICE OF RACK WIDTH

Wider racks are recommended when rack components cannot be cooled because of insufficient air due to a smaller than needed air flow area. This is a useful corrective comment provided a wider rack is ordered. Space is seldom available in banks of racks to remove one rack and replace it with a wider one. To do this may require many racks to be moved and several cables replaced as these must now be of longer or shorter length. Every rack moved, though only a few inches, will no longer have floor and rack openings in line. Electronics systems may have to be retested to confirm the soundness of new cables. A wider rack could be used to cool the overheating component but, to do so, dense packaging has to be avoided. Confirm space is available for the wider rack.

Racks are usually ordered before electronic designs are finalized. Replacing a rack can be costly because much electronics work must be undone to make needed corrections. The following analogous example is submitted to describe the nature of the problem.

Consider the chiller discussed in Section 19.9. In this example more water is needed on the lower floors than figured in the original design. Suppose the increase were 30%. This change will increase the line pressure loss by 1.30^2 and the HP required by 1.3^3 or 2.2 times. Thus a 25 HP motor must be replaced by a 60 HP motor. This is analogous to the increase in air flow needed for electronics overheating except that the quantity of air needed is likely to be doubled or more. The best solution is to increase the line size initially (translation: rack flow area).

Consider a relatively common rack size, 19 in. by 36 in., that is 80% blocked by electronic equipment. Its air flow area is $19 \times 36 \times 0.20 = 136.8$ sq in. Suppose the rack selected were 4 in. wider for the same rack equipment arranged the same way though on wider rack shelves. For this rack the added flow area is $4 \times 36 = 144$ sq in. Thus the air flow area is essentially doubled. In the example given in Section 30.3, Tight Packaging, a change such as this can double the air flow and halve the 60°F temperature thus eliminating the overheating problem.

30.5 SPACING BETWEEN RACKS

In the previous example a 4 in. wider rack will not solve the overheating problem unless there is 4 in. of floor space available within a bank of racks for the larger rack to be installed. To have space available a good way of proceeding is to leave 4 to 6 in. of floor space somewhere near the middle of a bank of racks. This should minimize the number of racks that may have to be moved should a wider rack be needed. One can get lucky and have the end rack be the wider rack needed. In any event the best solution is to choose the correct rack width in the first place. Be wary of a narrow rack that has a high cooling load and/or contains many components.

30.6 THE SCREEN ROOM DUCT CONNECTION

Screen rooms are shown in Figures 19-2 to 19-5. One that did not perform is shown in Figure 19-2 and discussed in Section 19.3. A frequently occurring problem in screen room designs that have external ducts is that the inlet elbow to the space beneath the screen room floor is too small. This elbow is not the same as a typical duct elbow. The construction needed may not be possible because of screen room details. The 10 in. floor is usually 10 in. above grade and 2 in. thick. A 2 in. lip is needed to make the elbow/screen room connection. Thus the actual opening is 6 in. high. Screen room vertical beams limit the elbow width to about 34 in. In some cases these limits have reduced the available flow area by 40%. The problem can be solved by raising the floor height to 15 in. or more substantially increasing the exit area of the elbow. The fix is simple if made before the floor is installed. When the small entrance area prevents adequate air flow, electronic changes may be necessary including assembly and disassembly. This is costly. The equivalent diameter of the space beneath the screen room floor is also a factor. In calculating it do not forget to include cables and other equipment installed there and that impede flow. Do not allow cables to be placed near inlets and outlets. At one installation a section of flooring (3 × 3, or 9 tiles total) was tilted upward to make room for a larger duct connection. The tilted screen room flooring was useable space lost in the screen room, but the design solved the elbow connection problem. Without it flow would be reduced as well as A/C capacity.

Electronic and project engineers take note. These conditions (adequate entrance and floor height) occurring in the screen room (acting as a duct) are your responsibility. The A/C manufacturer has no control over screen rooms. Heat transfer engineers, called after the fact, will not have a reasonable low-cost solution to this problem.

The above can be avoided if the A/C system is in the screen room and the return is along the floor inside the room (See Figure 19-4). The disadvantage is that the ECS occupies valuable screen room space and the air flow pattern is no longer vertical in the strict sense.

30.7 THE SIZE OF SCREEN ROOM INLET AND OUTLET DUCTS

The inlet to the space beneath the screen room floor has a high pressure loss in many installations because of construction needs at this connection. The same can be said of room exits. Grilles are often placed at exits for aesthetic reasons. These are relatively high pressure drop items. The operating properties at these connections are the kind that add to system pressure loss reducing the amount of cooling available (Section 19.3).

Suppose the total cooling required is 80,000 B/hr. The manufacturer's next larger size standard unit is chosen, in this case 84,000 B/hr (7 tons). An unintended way of reducing the cooling available is to have a higher than design pressure loss in the as-built duct system. Take note that nearly every subject covered in Section 19.3 adds to system pressure loss. Any could be the cause of lower than expected available cooling.

If a system is not performing the cooling required should be recalculated for every rack and room and totaled. Compare this total with the size of unit installed. The author's experience is that about one in five installations have undersized A/C or ECS units.

30.8 THE FLOW AREAS OF RACKS AND EXCHANGERS COMPARED

Electronics cooling differs from that of conventional exchangers. These should be understood to simplify problems that occur. The comparisons apply primarily to the approximately one in seven racks where electronics overheat. A shell-and-tube exchanger (the comparison applies to any exchanger) will be used for illustration. Suppose the initial selection has a calculated shell-side pressure drop that is higher than allowed. It can be reduced by increasing the baffle spacing, baffle cut, using a wider pitch, square pitch, increasing the shell OD, or reducing the number of tubes crossed.

Similar variations apply to cooling electronics but most are not available in the design of racks. Spacing between tubes is analogous to the spacing between components. Rack shelf space available often limits an increase in spacing between components. A wider rack could be selected (comparable to using a larger diameter shell) but this will not be an option unless space is available to install it in a bank of racks. The cost of this change (time, downtime, money) becomes a major factor.

Baffles are not needed in exchangers with short tube lengths. Electronic components are similar to baffles in that they direct flow but cannot be altered as baffles can. In most installations the only way of increasing flow area is to use a wider rack. The rack needed is difficult to identify and isolate from the six of seven racks that will not experience the problem.

If the inlet nozzle of a shell-and-tube exchanger is small, it will not allow the design quantity of fluid needed to enter. If a rack inlet is too small, flow will be limited. Undersized nozzles in electronics are, unfortunately, part of highly engineered components called racks (screen rooms also) which cannot be easily changed. The steps mentioned can be used to make a good shell-and-tube exchanger selection. A screen room that has adequate inlet and outlet connections plus floor height minimizes electronics overheating problems.

There are exceptions. If a wider rack is placed in service and "tight packaging" (Section 30.3) is employed, the overheating problem will not be solved because large flow areas are needed, the opposite of blockage that comes with tight packaging. The shell-and-tube exchanger mentioned had options available to reduce pressure drop. In electronics cooling systems, the problems are usually not part of the exchanger but of components in the cooling system, mainly screen room and racks. The latter are not usually designed for cooling as are exchangers.

The goal is to avoid electronics overheating on new installations. This is different from the recommendations needed to correct problems in existing racks. Some existing racks cannot be made wider and some screen rooms cannot have their floor heights raised because of lack of space. The solutions to these are approached differently.

30.9 THE SIZING OF RACKS AS DUCTS

A better way is needed for selecting racks so as to avoid electronics cooling problems. A rack that is too narrow (most racks in a system have the same depth) is the equivalent of providing an undersized duct in an A/C system. The rack selection is critical if electronics failures are to be avoided.

Existing field problems are different because the rack may be undersized, built, and in place. A correct initial selection will minimize overheating problems because it will have more air flow area and allow the components to be spaced further apart. A daunting problem is deciding which rack(s) should be wider.

The selection of racks is an early decision in the design of electronics systems. Racks must be on hand to engineer the spacing of the components that will be placed in them. The size to order is often an engineer's best judgment. When racks are selected the electronics to be placed in them are not completely known. The current method of selecting a rack's size is flawed and is one of the main reason for electronics overheating and system failures. In the past the assumption appears to have been that a rack can serve as a perfectly sized duct under all flow conditions, cooling loads, and component spacings, and at the same time, can have blowers added that will be the solution to every electronics cooling problem. Some racks, acting as ducts, are too narrow and cannot provide sufficient cool air to keep electronics cool. Most overheating problems can be solved by selecting a wider rack for the one in which the component failed.

A review of duct principles is in order. When ducts are designed, the area available for air flow is the product of the internal dimensions. Suppose 90% of this area is blocked. For this condition (it also depends on the amount of cooling required), about 10% of the air that could pass through the reduced opening will pass. In this case the air that can pass is probably less than is needed to provide the cooling sought. Conditions similar to this (minimum air flow areas) exist for many electronics units to be cooled and this condition is caused mainly by tight component spacing. Sometimes it is not one component but a bank of components that minimizes the air flow area. When less cooling air than desired is supplied, the result in many cases is component overheating.

The author has faced many overheating problems and in nearly every one the cause was minimum flow area which reduced air flow. Wider racks and component spacing (air flow areas) is the solution because this allows more cool air to pass. It is unfortunate that when this problem occurs, space is seldom available in banks of racks to replace an existing rack with a wider one. Hence, the source of the problem is the width of the initial rack selected assuming all racks have the same depth.

In the rack that has an overheating component it is unlikely the remaining components can be arranged to satisfy cooling needs without a study of heat transfer conditions. Remember, in packaging, the goal is greater air flow area, not greater air flow blockage. Another consideration is to place the item with the highest cooling requirement in the coldest air, that is, near the rack inlet. The following condition should be noted. After balancing the heat load, selecting and positioning blowers, installing baffles, etc., electronics engineers decide, at the last minute to add additional electronic boxes in the rack and place them in spaces intended for air flow. This is almost a guarantee that overheating will occur and is the kind of condition that should be considered before a rack order is placed. Avoid this condition. Last minute changes are seldom acceptable in any discipline without engineering approval. No solution to this problem is known to the author other than selecting a wider rack.

Not every rack must be wider, usually only one or two per installation. In reviewing the width needed, the following recommendations are made:

1. Similar applications with past overheating problems should be reviewed to see if a wider rack is apt to be needed before the selection is finalized.
2. Review racks that contain a large number of components as all block flow to some extent. In this case a wider rack may be needed.
3. A wider rack may be necessary because newer models of electronic devices may have larger face areas than older units.
4. Estimate the cooling needed in every rack. The greater this figure, the greater the chance that the rack should be wider to allow more air to pass.
5. Confirm that as more air is required, the racks are furnished with larger inlet and outlet openings.
6. If a wider rack is ordered, do not under any circumstances install more equipment on a given shelf. Remember, the goal is more air flow area, not flow blockage.
7. Leave floor space for a wider rack that may be needed later.

30.10 OVERVIEW OF ELECTRONICS COOLING ISSUES

The goal of heat exchanger and electronic engineers is to develop the best electronics cooling system possible. Human conditions surface when equipment fails and the question becomes, Who is responsible? A better understanding of the issues should help minimize confrontations between these disciplines.

In electronics cooling two pieces of equipment, screen rooms and racks, are often part of the duct system. Consider this: A/C and ECS manufacturers do not control either of these pieces of equipment. They have no voice in their design or selection. An undersized flow area in these installations is often the cause of an underperforming A/C system or of overheating rack components.

30.11 MOBILE EQUIPMENT

Many of the same problems experienced with racks and screen rooms can also be encountered with mobile equipment:

1. Ducts provided are too small.
2. Supply and return flow areas are too small.

3. Room airflow is blocked by other equipment (see Figure 19-1, example 4, and Section 24.6).
4. To save space, components are packed on tight spacing.

30.12 MAINTENANCE

Maintenance is one of the areas in which anything short of complete and accurate communication can have serious consequences. It has been noted repeatedly in this book that system pressure loss can be increased when maintenance personnel change or upgrade air filters without making these seemingly routine activities know to plant engineers.

Means for maintaining ECS and A/C units must be provided. The extent of this task is difficult to define, but if improperly done, it can result in a long-term shutdown of the electronic cooling system. Further, these problems are far more challenging than most electronic cooling problems. A rep-

Table 30-1 Maintenance Factors to Consider

Access to controls. To perform routine equipment checks, easy access to controls should be provided.

Access to site. A path must be available to move the unit to the installation site. In this regard, excess lift charges should be avoided.

Air filters. Filters should be replaceable without having to move entire units.

Compressors. The following compressor conditions should be satisfied. Access for routine maintenance of the compressor should be provided. In mobile vehicle applications, make sure the unit selected is designed for over-the-road service. In all units, built-in heaters should be easily accessible for routine checks of electrical continuity. Heaters must perform when needed. If the compressor space is minimal, make sure that there is an alternate design of compressor available that will satisfy dimensional needs, should it be needed. This limiting condition occurs more often than one would think.

Equipment removal or installation. A way should be provided for the easy installation or removal of equipment. Generally speaking, this should not require the dismantling of other equipment to accomplish the task.

Humidifiers. Make sure a means is available for servicing humidifiers.

Routine maintenance. Space should be provided for routine servicing.

Freeze protection. This problem frequently occurs in all types of A/C and ECS installations. Provide for it.

Shutdown. Many problems can occur following system shutdown. These include freezing, condensation on interior walls, refrigerant migration, an increase in room relative humidity, a decrease in room relative humidity, or the build-up of mold, to name a few. Provide for these as necessary for your application.

Table 30-2 Some Topics from Part 1, "Process Exchangers," and Part 2, "Air-Conditioning Exchangers,"That Also Apply to Electronics Systems

Chapter or section	Title or topic
Section 1.9	"Contract Requirements between Buyer and Seller"
Chapter 12	"Freezing"
Section 15.1	"Quantity of Refrigerant in System"
Section 17.3	"Noise"
Section 18.10	"Radiant Energy Affecting a Roof or Canopy"
Section 18.12	"Cables and Circuit Breakers"
Figure 19-1	Various causes of failures, or why field trips are often necessary
Figure 19-2	Diagram of basic ECS duct system for screen room
Figure 19-3	Screen room arrangement in which cooling air enters through ceiling ducts and returns to ECS unit at floor level
Figure 19-4	Screen room arrangement in which cooling air from ECS unit enters through ceiling ducts
Figure 19-5	"Ductless" screen room arrangement
Section 19.2	"Obstruction or Lack of Obstruction to Flow"
Section 19.3	"Screen Room Ductwork"
Section 19.4	"Airflows in A/C and ECS Systems"
Section 19.9	"Chiller Plant"
Section 19.11	"Starving Fans"
Section 19.12	"Flow Restrictions"
Section 19.15.3	"Rack Drawers"
Section 19.15.4	"Rack Plate or Flow Channel"
Section 19.16	"Summary: Causes and Cost of Failures"
Section 20.2	"Starving Fans"
Section 20.3	"Building Air Recirculation"
Section 20.4	"Duct Design"
Chapter 22	"Relative Humidity"
Section 22.1	"Parked Trailer Operating Condition"
Section 22.2	"Auditorium"
Table 23-2	Some causes of A/C units not meeting performance
Chapter 24	"Underperformance in A/C Systems Due to Ducting Problems"
Chapter 25	"Other Conditions That Cause A/C Systems to Underperform"
Chapter 26	"Ways of Increasing the Cooling Capacity of a System"
Chapter 27	"Conditions That Can Make an A/C System Unacceptable"

resentative list of problems faced is given in Table 30-1. Most of these are difficult to resolve once the equipment is built.

30.13 SUMMARY

Most overheating of electronics problems are the result of duct conditions in screen rooms and racks. Most screen room problems occur in designs that are similar to that shown in Figure 19-2 and that have a low floor height. In this design only, a low floor height results in a small entrance area, a high pressure loss condition. Small exit nozzles and low floor heights limit the amount of cooling that can be delivered and/or transferred between rooms. This problem is best resolved by raising the floor height allowing the entering nozzle size and duct flow area to be increased. The return outlet(s) and return duct should also be increased in size. These steps allow for the efficient transfer of cooling between rooms.

Some racks have too small an air flow area effectively limiting cooling. Identifying this rack(s) is not easy. Generally speaking, trouble occurs in racks with many components, high rack cooling loads, or the sizes of components increase. Once a problem rack(s) is identified, space must be available for installing it. If this recommendation is implemented, the number of overheating problems should be reduced by 85%. This assumes the issues of tight packaging, recirculation, and starving fans have been resolved. For existing problems, consider the recommendations given in Tables 29-3 and 29-4.

30.14 OTHER FACTORS

There are other reasons why electronic systems fail. For a summary of these, see Table 29-1, 29-2, and 29-5. In addressing overheating problems, do not overlook some topics from Part 1, "Process Exchangers," that also apply to electronics systems (see Table 30-2).

Chapter 31

The Cause of Most Field Problems

The following is intended to provide guidance in identifying field problems. Whenever a heat exchanger or heat exchanger system is not performing, the cause must be identified before a solution can be proposed. Locating the cause is seldom easy, often requiring a great deal of time, more likely weeks than days, and more than management would like to provide. Problems are less frequent when plate or double-pipe exchangers are used probably because these are seldom used in low pressure gas applications such as air cooling. Here are time consuming situations. Consider a large shell-and-tube exchanger that is thought to be the cause of a problem. It may take eight hours to disassemble and longer if positioned in a difficult area to access. After disassembly one may learn it was not the cause of the problem. Duct problems may be difficult because components may not be easily accessible due to elevation or because they are mounted behind a wall or false ceiling. Two months can pass before the cause of a problem is identified including those occurring in screen rooms and racks serving as ducts.

The author's experience is that most non-performing units or trouble jobs are *flow area* related. Examples of *flow area* conditions that affect performance follow. Similar problems occur at about the same rate in process, A/C, and electronics applications. Before addressing *flow areas*, the following steps should be taken.

1. Confirm that the amount of heating or cooling needed can be provided by the existing equipment and that an undersized unit is not the cause of trouble.
2. Check the value of the Reynolds and Prandl numbers raised to a power to assure the decimal point is correctly placed. A decimal point error has caused more trouble than one would think.

31.1 *FLOW AREA* CONDITIONS THAT HAVE CAUSED TROUBLE

- A flow path width or height dimension was too narrow preventing small particles from passing. Eventually these built-up and blocked flow, quench oil cooler (Sections 15.2 and 19.6), and plate exchanger (Section 19.7). A different *flow area* shape could solve this problem.
- As-built units installed at the site sometimes have lesser *flow areas* than those used in the calculations (See Sections 19.3 and Figure 19-2).
- The *flow area* provided was too small in new or existing installations (See Sections 11.2 and 19.9 through 19.11).

31.2 *FLOW AREA* CONDITIONS THAT AFFECT PERFORMANCE

- If equipment is placed in the flow path, the *flow area* can or be reduced (Figure 19-1 point 2) or remain the same (Figure 19-1 point 4), but flow is reduced in either case.
- Sometimes air returns through a room with a louvered door. Replacing the door with a solid one reduces the cooling system output because solid doors effectively block the *flow area* (Section 24.7).
- Replacing an air filter with a more efficient one without increasing *flow area* will block flow to some extent (Section 24.5).
- Undersized nozzles and ducts have less than ideal *flow areas*. Under these conditions less flow results (Sections 11.2 and 19.9).
- Duct take-off *flow areas* reduces the flow at other outlets as take-offs add to outlet flow area (Section 24.8).
- The lack of a dome *and its flow area* increases the entering pressure loss, and may be the cause of pitting or poor distribution (Section 15.2).
- Headers that have too small an air *flow area* may cause poor distribution (Section 19.13).

31.3 PROVIDING MORE *FLOW AREA* THAN WANTED CAUSES PROBLEMS SIMILAR TO THESE

- A missing seal effectively increases *flow area* resulting in short circuiting and an underperforming unit (Section 19.13).
- A shell-and-tube baffle hole that is too large increases *flow area* and reduces the rate, U, because longitudinal flow increases and cross-flow decreases (Section 19.14).
- An undersized baffle OD creates a *flow area* that allows fluid to bypass heat transfer surface (Section 19.14).
- The lack of seal strips (increasing *flow area*) causes flow to bypass surface making part of the exchanger ineffective (Section 19.13).

31.4 DESIGN CONDITIONS

There are other conditions that affect *flow area* when selecting equipment. These include a water velocity limit, available space particularly for piperack mounted units, the aspect ratio of coils installed in ducts, fan coverage, simplified manifolding, shipping limitations, low cost offerings, pressure drop limits, whether an A/C (400 cfm) or ECS (550 cfm) unit is used, plus others. It should be noted that *flow areas* listed in this Section cannot always be separated into the categories given in this article. Conditions overlap.

Flow area needs listed under "Design Conditions" are usually factored into the selection by the rating engineer (specific needs should be in the specifications). Equipment is selected to meet these needs. These are not true field problems beyond the control of the heat exchanger engineer. Mainly these are undersized ducts, pipes, and approach flow areas provided by others.

The author's experience, as a user of equipment, is that two kinds of *flow area* problems frequently occur. One is that equipment, selected by others, is now used for purposes other than that for which they were designed. The second is that problems occur about as often in equipment associated with heat exchangers as in the heat exchanger itself. Here are examples.

Increased flow in chiller piping (Section 19.9).
Starving fans—Equipment installed too close to a building or wall (Section 19.11).
Screen room serving as duct—Low floor height and an undersized duct and rack system. (Section 19.3).
Tight packaging in racks (Section 30.3).
Flow path blocked
  Figure 19-1 point 2
  Figure 19-1 point 4
  Quench oil cooler (Section 19.6).
  Plate exchanger (Section 19.7).
Undersized header (Section 11.2).
Flow restrictions (Section 19.12).
Seal strips and bypass areas (Section 19.13).
Louvered door (Section 24.7).
Air filter (Section 24.5).

The point is that *flow area* is the most common cause of field problems. Our purpose is to identify problems so that solutions can be more readily attained.

Index

C

D